ENVIRONMENTAL/SAFETY REGULATORY COMPLIANCE FOR THE OIL & GAS INDUSTRY

ENVIRONMENTAL/SAFETY REGULATORY COMPLIANCE FOR THE OIL & GAS INDUSTRY

George H. Holliday, Ph.D., P.E., DEE
Holliday Environmental Services, Inc.
Tomball, Texas

PennWell Books

PennWell Publishing Company
Tulsa, Oklahoma

PennWell Publishing Company
1421 South Sheridan/P.O. Box 1260
Tulsa, Oklahoma 74101

Holliday, George H.
Environmental/safety regulatory compliance for the oil & gas Industry/George H. Holliday.
p. cm.
Includes Index.
ISBN 0-87814-452-8
1. Petroleum law and legislation—United States. 2. Gas—Law and legislation—United States. 3. Petroleum industry and trade—Safety regulations—United States. 4. Gas industry—Safety regulations—United States. 5. Environmental law—United States. I. Title.
KF 1849.H65 1995
343.73'0772—dc20
[347.303772] 95-23297
CIP

Printed in the United States of America

1 2 3 4 5 99 98 97 96 95

CONTENTS

TABLES

FIGURES

FORMS

Form		Page

ACRONYMS

mg/m_3	Microgram per cubic meter [ppm, weight/volume]
ACP	Area contingency plan
APA	*Administrative Procedure Act*
API	American Petroleum Institute
ARC	Agency for Reseach on Cancer
BAT	Best available control technology
bbl	Barrel [42 US gallons]
BCT	Best conventional control technology
BOD^5	Biochemical oxygen demand
BPJ	Best professional judgment technology
BPT	Best practicable control technology
BTEX	Volatile organic compounds benzene, toluene, ethylbenzene o-,m - and pxylenes
CAA	*Clean Air Act*
CERCLA	*Comprehensive Environmental Response, Compensation and Liability Act*
CFR	*Code of Federal Regulations*
CO	Carbon monoxide
COA	Corresponding onshore area
COD	Chemical oxygen demand
COTP	Captain of the port
CWA	*Clean Water Act*
DOI	Department of Interior
DOT	Department of Transportation
E&P	Exploration and production [a function of oil companies]
EHS	Extremely hazardous substance
EPA	Environmental Protection Agency
EPCRA	*Emergency Planning and Community Right-to-Know Act*
ERW	Electric resistance welded
FIFRA	*Federal Insecticide, Fungicide, and Rodenticide Act*
FP	Flash point
FR	*Federal Register*

GACT	Generally available control technology
H^2S	Hydrogen sulfide
HAP	Hazardous air pollutants
HAZCOMM	Hazard communication program
HAZWOPER	Hazardous waste operations and emergency response
HC	Hazardous chemical
ICE	Internal combustion engines
LC50	Lethal concentration
LD50	Lethal dose
LAER	Lowest achievable emissions rate
LEPC	Local emergency planning commitee
MACT	Maximum achievable control technology
MCL	Maximun contaminant level
mg/kg	Milligrams per kilograms; ppm by weight
mg/l	Milligrams per liter, ppm by weight and volume
mg/m^3	Milligram per cubic meter
MIT	Mechanical integrity test
ml/m^3	PPM by volume
MMS	Mineral Management Service [Dept. of Interior]
MMSCFD	Million standard cubic feet per day
MOU	Memorandum of understanding
MSDS	Material safety data sheet
MTR	Marine transportation-related
N	Nitrogen
NAAQS	National Ambient Air Quality Standards
NCP	National contingency plan
NESHAPS	National Emission Standards for Hazardous Air Pollutants
NFPA	National Fire Prevention Association
NOI	Notice of intent
Nox	Oxides of nitrogen
NPDES	National Pollutant Discharge Elimination System [Permit]
NSPS	New Source Performance Standard
O&G	Oil and grease
O_3	Ozone
OCS	Outer Continental Shelf

OPA-90	*Oil Pollution Act of 1990*
OSHA	Occupational Safety and Health Administration [or Act] [Dept. of Labor]
OVA	Organic vapor analyzer
P.E.	Professional engineer
Pb	Lead
PCB	Polychlorinated biphenyl
PEL	Permissible exposure limit
PL	Public Law
PM_{10}	Particulate matter having an aerodynamic diameter less than 10 microns
POTW	Publicly owned treatment works
ppm	Parts per million, typically volume/volume or weight/weight
PPP	Pollution prevention plan
PREP	National Preparedness for Response Exercise Program
PRP	Potenially responsible party
PSD	Prevention of significant deterioration of air quality
PWS	Public water system
r	Radius
RACT	Reasonably available control technology
RCRA	*Resource Conservation and Recovery Act*
RQ	Reportable quantity
RSPA	Research and Special Projects Administration
SARA	*Superfund Amendments and Reauthorization Act*
SDWA	*Safe Drinking Water Act*
SERC	State emergency response commission
SIC	Standard industrial classification
SIP	State implementation plan
SLA	*Submerged Lands Act*
SMYS	Specified minimum yield strength
SO_2	Sulfur dioxide
SPCC	Spill prevention, control and counter measure
STEL	Short-term exposure limit
TCLP	Toxicity characteristic leaching procedure
TDS	Total dissolved solids

TLV	Theshold limit value
TOC	Total organic carbon
TPQ	Threshold planning quantity
TPY	Tons per year
TSCA	*Toxic Substance Control Act*
TSD	Treatment, storage and disposal
TSS	Total suspended solids
TWA	Time weighted average
UIC	Underground injection control
UN	United Nations
USC	U.S. Code
USCG	Coast Guard [Dept. of Interior]
USDW	Underground source of drinking water
VOC	Volatile organic compound
VRU	Vapor recovery unit

TERMS

≥ greater than or equal to
≤ less than or equal to
< less than
> greater or more than
Attainment area — a geographical area where the National Ambient Air Quality Standards are not exceeded
Countermeasure — the act of remediating a chemical or oil spill
Criteria pollutants — nitrogen oxides, sulfur dioxide, carbon monoxide, ozone, volatile organic compounds and hydrocarbons, particulate matter, and lead.
Effluent — liquid discharge
Emission — gaseous release
Fugitive emission — unintentional gaseous release from sealing member, i.e., flanges, couplings, seals, etc.
Gray water — domestic waste water
Hazardous air pollutant — 189 air pollutants identified in the *Clean Air Act Amendments of 1990*
National Ambient Air Quality Standards — concentration of air pollutants above which air is considered unhealthful
Non-attainment area — a geographical area where the National Ambient Air Quality Standards are exceeded
Operating permit — a required air permit issued if the equipment emits 100 TPY of a criteria pollutant or 10 TPY of a single hazardous air pollutant or 25 TPY of a combination of air pollutants
Polychlorinated biphenyl — a biphenol atom to which is attached more than one chlorine atom; very persistent compound
Release — leak of liquid or emission of gas into air, water or onto the soil
Significant deterioration — an increase in criteria pollutants causing an exceedance of the National Primary Air Quality Standards
Volatile organic compounds — non-methane, non-ethane hydrocarbons

PREFACE

This environmental compliance handbook is written for the nonlawyer. This is not to say lawyers cannot find valuable and useful information in the handbook. However, the target readers are individuals interested in ascertaining environmental compliance information with U.S. Environmental Protection Agency regulations without the legal phraseology and mumbo jumbo.

Such an audience is broad in scope and generally not familiar with regulations. To serve these readers, I have attempted to place all the necessary tools in this handbook. Accordingly, the reader will find liberal use of citations to the *Federal Register, Code of Federal Regulations,* and the U.S. Code.

These government and private publications typically are available at public, state, and federal libraries. The reader will find these publications useful in confirming the interpretations contained in this handbook or in reveiwing the published rules.

The method of citation for each of the documents is similar but varies in detail.

The *Federal Register* is a daily publication of the U.S. Government. Some states publish their own registers, i.e., *Texas Register.*

Except in rare cases, this book concentrates on federal requirements.

These registers provide copies of all notices, corrections, proposed rules, and final rules promulgated by branches of federal or state governments. The *Federal Register* is divided into various parts called Titles. There are 50 Titles. However, this book deals with:

- Title 29, Occupational Safety and Health [OSHA]
- Title 30, Mineral Management Services [MMS]
- Title 33, Coast Guard [USCG]
- Title 40, Environmental Protection Agency [EPA]
- Title 43, Bureau of Land Management [BLM]
- Title 49, Research and Special Programs Administration [RSPA]

In all of these cases, the *Federal Register* citation will consist of a volume number [56], *Federal Register* [FR], and page number [16274]. Thus, the citation will be 56 FR 16274. It is complete as given, however, it is more convenient if the publication date of the *Federal*

Register is included. Wherever possible, this book includes the publication date when citing the *Federal Register*.

The *Code of Federal Regulations* [CFR] is the collection of all final rules promulgated under the *U.S. Administrative Procedure Act* [APA].

The CFR editions are published annually but are not timely, so they are not always current. Therefore, it is prudent to review the *Federal Register* to make certain no revisions to the rule have been published since the CFR was published. The *Code of Federal Regulations* is cited by Title number [40], *Code of Federal Regulations* [CFR], and part number [Part 261] or section number [§261.3], i.e., 40 CFR Part 261 or 40 CFR §261.3.

The U.S. Code [USC] is a privately published series of books containing the statutes as well as the court decisions and interpretations regarding those statutes. The citations include subject number [5], U.S. Code [USC], and section number [552], i.e., 5 USC 552.

I have taken the liberty of interpreting the regulations based on extensive experience and review. I believe the interpretations are correct and usable. However, interpretations are in the "eye of the beholder." Thus, reasonable men and women can differ in the interpretation of the same statute or rules. I recommend using the interpretations as provided, *but be aware the agency may interpret the regulation differently.*

This handbook is divided into chapters by federal statutes. I have attempted to make each chapter stand alone to avoid repeated reference to another chapter.

I hope the reader finds the handbook useful.

George Holliday
Tomball, TX

INTRODUCTION

This is an overview of each federal environmental/safety statute. I believe the reader will benefit from the overview, because knowing which statute applies to the particular situation eliminates searching the maze of statutes for appropriate information. Further, the indexes to this book are arranged by the federal statutes to encourage efficient use of the book.

CLEAN AIR ACT

The *Clean Air Act* [CAA] mandates control of air contaminants from combustion sources [stationary and mobile], fugitive emissions [criteria and hazardous air pollutants], glycol dehydrator emissions [hazardous air pollutants] and sulfur plant emissions [SO_2].

Additionally, the CAA requires operating permits for stationary sources of criteria air pollutants at or above 100 tons per year [TPY] or Hazardous Air Pollutants [HAP] at or above 10 TPY for any one HAP or a combination of HAP at or above 25 TPYs.

CLEAN WATER ACT

The *Clean Water Act* [CWA] mandates reporting any release of oil into Waters of the U.S. Also, the CWA requires permits for discharges of contaminated process water or storm water into Waters of the U.S. The operator of a permitted storm water facility must prepare and implement a pollution prevention plan for the facility.

Further, CWA regulations require preparation of a spill prevention control and countermeasure [SPCC] plan for all facilities storing more than 660 gallons of oil in a single tank or 1,320 gallons of oil in multiple tanks.

OIL POLLUTION ACT OF 1990

The Oil Pollution Act of 1990 [OPA-90] is an extension of the CWA. The act mandates preparation, submission, and approval of a response plan, if hydrocarbons are stored in excess of threshold quantities. Jurisdiction of the act rests with four agencies:

- Environmental Protection Agency – onshore and certain offshore non-transportation-related facilities [tanks and over-water transfers]
- Coast Guard – offshore marine transportation-related facilities [ship and loading facilities]
- Mineral Management Service – offshore non-transportation facilities [platforms]
- Research and Special Programs Administration – onshore and certain offshore pipelines

A Memorandum of Understanding delineates each agency's jurisdiction. The response plan requires extensive planning, training, and recordkeeping.

SAFE DRINKING WATER ACT

The *Safe Drinking Water Act* [SDWA] imposes controls on water systems serving the public or company personnel in excess of a threshold number of people. Onsite company water wells, whether controlled by the regulations or not, should adhere to the sampling and analytical provisions of the act to avoid exposing employees to unhealthful water. Also, the SDWA mandates permits for five classes of injection wells. Class II injection wells include disposal and enhanced recovery injection wells. The permit requires specified well completions, monitoring, recordkeeping, and integrity testing.

TOXIC SUBSTANCES CONTROL ACT

The *Toxic Substances Control Act* [TSCA] controls use, storage, and disposal of polychlorinated biphenyls, and the disposal of asbestos. Additionally, the statute requires:

- reporting of manufactured chemicals [ethane, propane, n-butane, iso-butane, pentane and raw gasoline mixtures]
- use of any chemicals included on the master inventory

- recording of allegations
- reporting of health and safety data
- reporting of substantial risk

RESOURCE CONSERVATION AND RECOVERY ACT

The *Resource Conservation and Recovery Act* [RCRA] controls storage and disposal of solid waste. Solid waste includes non-hazardous and hazardous wastes. Certain exemptions allow Exploration and Production [E&P] operators latitude in disposing of typical oilfield wastes. This includes recycled waste, i.e., used lubricating oil, batteries, and chemicals. The definitions of exempted waste are specific and provide little tolerance.

COMPREHENSIVE ENVIRONMENTAL RESPONSE, COMPENSATION, AND LIABILITY ACT

The *Comprehensive Environmental Response, Compensation and Liability Act* [CERCLA] provides clean-up requirements and funds for orphaned hazardous waste sites. Also called Superfund, CERCLA can and has been applied to non-hazardous wastes where the waste imposes an "imminent and substantial threat of endangerment to human health and the environment," i.e., produced water or condensate pits contaminating ground water.

Additionally, CERCLA requires reporting of certain releases of hazardous substances [treating chemicals, acids] to the environment, i.e., air, water or soil.

HAZARD COMMUNICATION STANDARD

The Hazard Communication Standard [HAZCOMM] is mandated by the *Occupational Safety and Health Act* [OSHA]. Among many other things, OSHA requires a program to communicate health and safety hazards of chemicals to which an employee is exposed. Hazardous chemicals include all substances imposing physical and health impacts on employees. Such innocuous chemicals as reproduction and printer toners are classed as hazardous chemicals. Any employer exposing an employee to any quantity of a hazardous chemical must:

- prepare a written HAZCOMM program
- train employees in the written program
- prepare a list of hazardous chemicals present at the workplace
- maintain current material safety data sheets
- label all hazardous chemicals in the workplace

SUPERFUND AMENDMENTS AND REAUTHORIZATION ACT

The *Superfund Amendments and Reauthorization Act* [SARA] mandates notifying the state, local agency [county or parish] and fire department of the presence of 10,000 lb or more of a hazardous chemical or the threshold planning quantity of an extremely hazardous substance. Additionally, SARA requires submitting an annual inventory of these chemicals on or before March 1st of each year.

1 Clean Air Act

SUMMARY

The *Clean Air Act*[1] and particularly the *Clean Air Act Amendments of 1990* have the potential to financially damage all U.S. Exploration and Production [E&P] operations, both mobile and stationary. The 1977 *Clean Air Act* [CAA] exempted all *existing* major stationary sources from construction permitting. The 1990 amendments preserved the exemption but imposed operating permit requirements on all sources of criteria pollutants [NO_x, SO_2, CO, PM_{10} and O_3] at or above:

- 100 TPY in attainment areas
- major source threshold in non-attainment areas

Also, the amendments impose operating permit requirements and maximum achievable control technology [MACT] on major sources of HAP ≥10 TPY for any single HAP or 25 TPY for any combination of HAP. HAP of interest to E&P operators are benzene, toluene, ethylbenzene, xylenes [BTEX] and formaldehyde. This provision affects glycol dehydrators having throughput of about 5-10 million standard cubic feet per day [MMSCFD] and higher and may affect large engine installations.

[1] 42 USC 7401 et seq.

INTRODUCTION

This chapter presents provisions of the CAA and recent amendments, particularly, relating to major stationary source permitting, mobile sources, HAP, operating permits and New Source Performance Standards [NSPS]. Also, suggestions are offered for early action by operators to avoid or confirm applicability of the CAA to E&P facilities.

The CAA was passed as Public Law [PL] 88-206 on December 17, 1963. Congress amended the act numerous times. The most significant amendments occurred on August 7, 1977, as the *Clean Air Act Amendments of 1977*[2] and on November 15, 1990, as the *Clean Air Act Amendments of 1990*.[3] In its present form, the CAA represents the most pervasive and intrusive environmental statute ever enacted in the United States.

The statute is divided into 11 Titles. Titles I, II, III, and V directly impact E&P operators. These deal with:

- air pollution prevention and control
- emissions standards for moving sources
- general
- permits

AIR POLLUTANTS

The CAA mandates control of two types of air pollutants:

- criteria pollutants
- hazardous air pollutants

CRITERIA POLLUTANTS

The criteria pollutants consist of:

- nitrogen oxides [NO_x]
- sulfur dioxide [SO_2]

---

2 PL 95-95

3 PL 101-549

- particulate matter having an aerodynamic diameter ≤10 micrometers [PM_{10}]
- carbon monoxide [CO]
- ozone [O_3]
- lead [Pb]

The degree of control of criteria pollutants depends on the concentration of criteria pollutants in the ambient air of the area of interest, as discussed below.

HAZARDOUS AIR POLLUTANTS

HAP consists of 189 statutorially established air toxics. Control depends only on the quantity of air toxics emitted.

NATIONAL AMBIENT AIR QUALITY STANDARDS

The acceptable levels of air quality are known as the National Ambient Air Quality Standards.[4] The National Ambient Air Quality Standards [NAAQS] are divided into primary and secondary ambient air quality standards.

The *primary standards* define levels of air quality judged necessary to protect public health. The *secondary standards* define levels of air quality judged necessary to protect public welfare. These standards apply to the criteria pollutants.

The NAAQS were obtained from available health and welfare literature,[5] Table 1–1.

AIR QUALITY DESIGNATION AREAS — CRITERIA POLLUTANTS

The CAA divides the United States into three areas in accordance with the air quality within each area. The degree of emissions control for criteria pollutants depends on the air quality designation area.

---

[4] 40 CFRPart 50

[5] 40 CFR §§50.4 – 50.12

Table 1–1 National Ambient Air Quality Standards for criteria pollutants

Criteria Pollutant	NAAQS Primary (μg/m³)	NAAQS Secondary (μg/m³)
NO_x — Annual arithmetic mean	100	100
SO_2 — Annual arithmetic mean	80	—
Max. 24-hr. concentration not to be exceeded >1 time per year	365	—
Max. 3-hr. concentration not to be exceeded > 1 time per year		1300
PM_{10} — 24-hr. average concentration	150	150
Annual arithmetic mean	50	50
CO — 8-hr. average concentration not to be exceeded >1 time per year	10	—
1- hr. average concentration not to be exceeded >1 time per year	40	—
O_3 — Max. hourly average concentration not to be exceeded 1 time per year	235	235
Pb — Max. arithmetic mean averaged over a calendar quarter	1.5	1.5

Attainment Areas – Criteria Pollutants

Any geographic area wherein the ambient air concentration of a criteria pollutant is equal to or less than the primary NAAQS for that pollutant is an attainment area for that pollutant.[6] Each criteria pollutant is considered separately when determining area designations. Accordingly, an area can attain the NAAQS for one or more criteria pollutants and exceed the NAAQS for other criteria pollutants. Most major oil producing provinces are within attainment areas except California.

Non-Attainment Areas – Criteria Pollutants

Any geographic area wherein the ambient air concentration of a criteria pollutant exceeds the primary NAAQS for that pollutant is a non-attainment area for that pollutant. Each criteria pollutant is considered separately when determining area designations. Accordingly, a location can be a non-attainment area for one or more pollutants and an attainment area for the others.

The *Clean Air Act Amendments of 1990* classified ozone non-attainment areas by the severity of non-attainment, Table 1–2. Also, the amendments imposed deadlines by which the classified non-attainment areas must achieve attainment status,[7] Table 1–2.

Table 1–2 Ozone non-attainment area classifications

Area Class	Concentration*	Primary Standard Attainment Date**
Marginal	0.121 up to 0.138	3 years after Nov. 15, 1990
Moderate	0.138 up to 0.160	6 years after Nov. 15, 1990
Serious	0.160 up to 0.180	9 years after Nov. 15, 1990
Severe	0.180 up to 0.280	15 years after Nov. 15, 1990
Extreme	0.280 and above	20 years after Nov. 15, 1990

* The concentration value is measured in parts per million (ppm).
** The primary standard attainment date is measured from November 15, 1990.

6 42 USC 7407 (9) (1) (c)
7 42 USC 7511 (a) (1)

Further, the amendments specify the emission rate threshold and the offset ratios for each non-attainment area classification,[8] Table 1–3.

Table 1–3 Major source definitions and offset ratios in ozone non-attainment areas

Category	Size of Major Source	Offset Ratios
Marginal	100 TPY	1.10:1
Moderate	100 TPY	1.15:1
Serious	50 TPY	1.20:1
Severe	25 TPY	1.30:1
Extreme	10 TPY	1.50:1

States have the option of choosing a major source definition of 5 TPY (and accepting other conditions) in order to avoid complying with the requirement that emissions be reduced 15% over the first six years.[9]

Few producing provinces are in non-attainment areas such as in California.

Unclassified Areas – Criteria Pollutants

Attainment and non-attainment areas are designated on the basis of ambient air quality measurements. Some areas, primarily non-industrialized areas, cannot be classified, because no ambient air monitoring data are available. These are known as unclassified areas. Any major stationary source built in an unclassified area must conduct pre-permitting ambient air quality monitoring for the proposed criteria to be emitted. These data are used to classify the area, as either attainment or non-attainment.

[8] 42 USC 7511 (a)

[9] 42 USC 7512 (b) (1) (ii)

MAJOR STATIONARY SOURCES — CRITERIA POLLUTANTS

The threshold for a major stationary source of criteria pollutants depends on whether the area is designated as attainment or non-attainment.

Attainment Areas – Criteria Pollutants

An emitting source of criteria pollutants within an attainment area becomes a "major stationary source" for that criteria pollutant when it emits > 250 TPY,[10] except any of the following facilities become major stationary sources at 100 TPY.[11]

- fossil fuel-fired steam electric plants of > 250 million BTU/hr. heat input
- Coal cleaning plants (with thermal dryers)
- kraft pulp mills
- Portland cement plants
- primary zinc smelters
- iron and steel mills
- primary aluminum ore reduction plants
- primary copper smelters
- municipal incinerators capable of charging > 250 tons of refuse per day
- hydrofluoric, sulfuric, and nitric acid plants
- petroleum refineries
- lime plants
- phosphate rock processing plants
- coke oven batteries
- sulfur recovery plants
- carbon black plants [furnace process]
- primary lead smelters
- fuel conversion plants
- sintering plants
- secondary metal production plants

---

[10] 42 USC 7491 (g) (7)

[11] 40 CFR §52.21

- chemical process plants
- fossil fuel boilers or combinations thereof totaling more than 250 million BTU/hr. heat input
- petroleum storage and transfer units with a total storage capacity exceeding 300,000 bbl
- taconite ore processing plants
- glass fiber processing plants
- charcoal production plants

All major stationary sources in attainment areas must apply for and be granted a Prevention of Significant Air Quality Deterioration [PSD] permit before commencing construction of a new or modified facility.[12]

A modified source refers to a major modification or a modification to a non-major source which makes the modified source a major stationary source.[13] If the source is a major stationary source, a major modification is a physical or operational change resulting in a significant net emission increase of a criteria pollutant.[14] The following levels are considered significant increases:[15]

- CO = 100 TPY
- NO_x = 40 TPY
- SO_2 = 40 TPY
- ozone = 40 TPY of volatile organic compounds
- lead = 0.6 TPY

The CAA prohibits significant deterioration of air quality at the proposed facility site or modification site. However, the act allows for consumption of an "increment of progress." The applicant for a PSD permit must demonstrate the facility emissions do not cause an exceedance of the increment.[16] In practice, the task is more rigorous, because each applicant is assigned a portion of the increment. The permit applicant for a new or modified facility must demonstrate the facility does not consume more than the assigned portion of the increment. This procedure allows all applicants an opportunity to construct a new facility or modify an existing facility in an attainment area.

---

12 40 CFR §51.166
13 40 CFR §51.166 (b) (23)
14 40 CFR §51.165 (a) (1) (v)
15 40 CFR §51.165 (a) (1) (x)
16 40 CFR §55.6

Conceivably, the increment eventually may be consumed. In that event, each existing permitted facility must reduce emissions by installing emissions controls to make a portion of the increment available for other applicants.

Reduction of emissions from proposed new or modified facilities are mandated by use of the best available control technology [BAT][17] or New Source Performance Standards [NSPS].[18]

BAT constantly changes because of industrial progress in emissions control. Accordingly, permit provisions become more severe with time.

Non-Attainment Areas – Criteria Pollutants

The threshold emissions rates for a major stationary source in a non-attainment area were revised by the *Clean Air Act Amendments of 1990* and depend on the category of non-attainment for ozone [volatile organic compounds (VOC)], Table 1–2.

The threshold ranges from 100 TPY for moderate and marginal areas to 10 TPY for extreme areas, Table 1–3.

All major stationary sources in a non-attainment area must apply for and be granted a new source review permit before construction of a new or modified facility.[19] The owner of a new facility must apply lowest achievable emission rate [LAER] and NSPS, if applicable. Further, the owner must provide offsets of at least 1:1 for all criteria pollutants for which the area is classed non-attainment, except areas classed non-attainment for ozone where offsets depend on the non-attainment category, Table 1–3.

Existing VOC [non-methane, non-ethane hydrocarbons] or NO_x sources in ozone non-attainment areas must apply reasonably available control technology [RACT].[20] RACT requirements probably will be more severe than LAER requirements and will be imposed on existing sources that previously have applied LAER. Exactly how RACT will be applied will depend on the State Implementation Plan [SIP].

The intent of the non-attainment control technology is to force states containing non-attainment areas to make reasonable progress toward attainment.

---

17 42 USC 7475 (a) (4)

18 42 USC 7411 and 40 CFR Part 60

19 40 CFR §51.165 (a) (1) (xvi)

20 42 USC 172 (b) (2)

MOBILE SOURCES — CRITERIA POLLUTANTS

Mobile Sources are controlled under *Clean Air Act Amendments of 1990* Title II.[21] This title also deals with heavy engines [drilling and well pulling] and vehicular engines.

Heavy-Duty Engines

Starting with model year 1998, new heavy-duty gasoline and diesel engines for mobile sources, trucks, and pulling units may not exceed 4 grams per brake horsepower hour.[22]

The CAA amendments provide for permitting of temporary sources.[23] Controls are not mandatory, and there are no implementing regulations. However, it is possible states may elect to impose permitting requirements on well pulling machines and onshore drilling rigs. Though this is doubtful, operators must remain vigilant to the possibility.

Also, the EPA is required to study the feasibility of controlling rebuilt engine emissions by regulation.[24]

Automobiles and Light-Duty Trucks

The CAA requires maintenance of factory-installed emissions controls.[25] Further, a fleet operator [a company operating 10 or more vehicles from a single location] in a severe or serious non-attainment area, Table 1–2, must convert to alternate fuel, i.e., natural gas, electricity, LPG. This provision typically does not impact small- to medium-size operators, because most company vehicles are located in field locations which are predominately in attainment areas. Similarly, the trip reduction provision [forced car or van pooling] of the CAA does not impact small- and medium-size operators, because they seldom have 100 employees in a serious or severe ozone non-attainment area.[26]

---

[21] 42 USC 7401

[22] 42 USC 7521 (a) (3) (B) (ii)

[23] 42 USC 7661c. (e)

[24] 42 USC 7521 (a) (3) (D)

[25] 42 USC 7541 (g)

[26] 58 FR 13596, 12 March 1993

HAZARDOUS AIR POLLUTANTS

Title III prescribes 189 HAP that must be controlled, Appendix 1.[27] HAP of greatest interest to oil and gas operators are BTEX and formaldehyde.

H_2S is not a hazardous air pollutant for purposes of the act.

The CAA mandates a new approach to controlling HAP. In essence, the mandate calls for "shooting first and asking questions afterwards." First, it requires application of maximum achievable control technology [MACT] to all major sources of hazardous air pollutants.[28] Then, after application of MACT, the statute requires a risk assessment to determine if MACT was adequate, and if not, additional controls are required.[29]

SOURCE TYPES — HAZARDOUS AIR POLLUTANTS

Two types of HAP sources are specified by the CAA:

- major sources[30]
- area sources[31]

Major Source – Hazardous Air Pollutants

For purposes of the *Clean Air Act Amendments of 1990,* a major source of HAP is any facility emitting ≥10 TPY of any single HAP or 25 TPY of any combination of HAP.[32] All other sources of HAP are area sources. Fugitive emissions associated with the source must be included in the calculation.[33] *E&P facilities are not aggregated to form major sources.*[34] This means the operator is not required to aggregate all sources of HAP on a lease or contiguous leases. Each tank battery,

---

[27] 42 USC 7412 (b)
[28] 42 USC 7412 (d) (3) (A)
[29] 42 USC 7412 (f)
[30] 42 USC 7412 (a) (1)
[31] 42 USC 7412 (a) (2)
[32] 42 USC 7412 (A) (11)
[33] 42 USC 7412 (d) (2)
[34] 42 USC 7412 (n) (4) (A)

compressor station, pump station, gas plant, etc. stands alone and is calculated separately. However, EPA has not promulgated regulations defining aggregated sources. "Wells and associated equipment" are discussed in the *Clean Air Act Amendments of 1990* in relation to aggregation.[35] Accordingly, EPA should interpret the aggregation exemption to apply to wells and the associated equipment, i.e., dehydrators.

The most likely major sources of HAP in E&P operations are glycol dehydrators. Based on current knowledge, *a glycol dehydrator having a throughput of 5-10 MMSCFD probably is a major source.* The actual emissions depend on the BTEX content of the inlet gas and the amount of water extracted. Thus, glycol dehydrators processing small volumes of rich gas may become major sources.

Two methods are available for determining HAP emissions from dehydrators:

- mass balance of rich and lean glycol flow within the dehydrator
- GRI-GLYCalc computer program for ethylene and triethylene glycol dehydrators

Two other methods are available for other E&P equipment:

- GRI-HAPCalc[36] computer program for amine units, storage tanks, reciprocating engines, combustion turbines, small combustion devices, and fugitive emissions.
- API/EPA AP-42 Factors[37] for fugitive emissions from onshore and offshore E&P operations.[38]

E&P facilities were included as a major source in the "Initial List of Categories" of HAP on July 16, 1992.[39]

---

[35] 42 USC 7412 (n) (4)

[36] Gas Research Institute Computer Program, Chicago, Illinois. Order from Radian Corporation, P.O. Box 2011088, Austin, TX 78720 at $25 per copy.

[37] EPA, *Compilation of Air Pollution Emission Factors, Vol. 1: Stationary Point and Area Sources*, AP-42, 4th ed., 1993.

[38] Factors based on total organic compounds use GRI speciation factors to determine BTEX.

[39] 57 FR 31576

Glycol Dehydrator Sampling Methods

The accuracy of the computer programs is dependent on proper sampling practices.

The best information comes from the Gas Research Institute. Their contractor, Radian Corporation, initially compared three methods:

- EPA modified compendium method TO-14,[40] Figure 1–1
- Gas Processor Association method 2166-86,[41] Figure 1–2
- total capture

Figure 1–1 Apparatus for use with modified EPA compendium sampling method TO-14

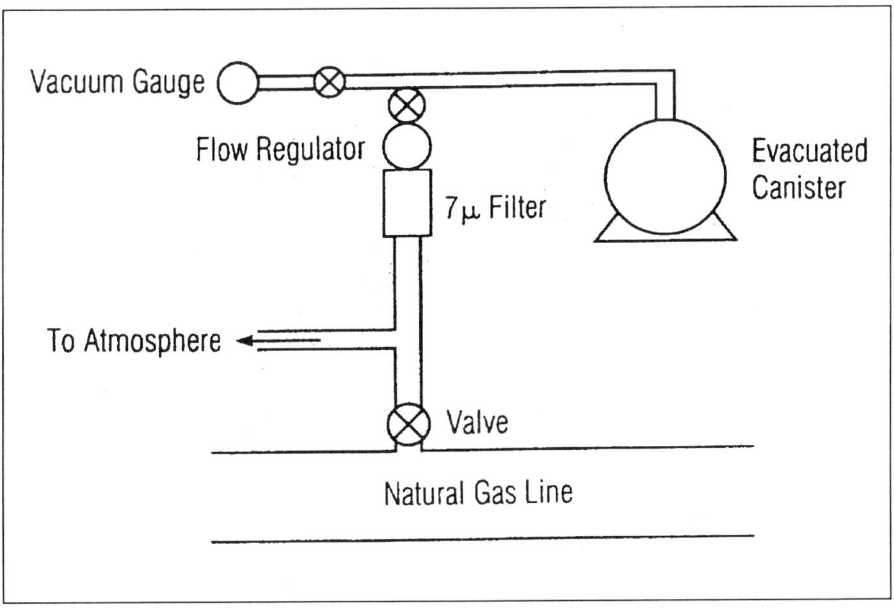

[40] Uses evacuated canister, collects sample at less than atmospheric pressure and uses gas chromatography with multiple sectors [GC-MD] analytical method.

[41] Uses evacuated or helium-filled canister, collects sample at line pressure and uses gas chromatography with thermal conductivity dectector and flame ionization dectector [GC-TCD/FID].

Figure 1–2 Apparatus for commonly used industry sampling approach

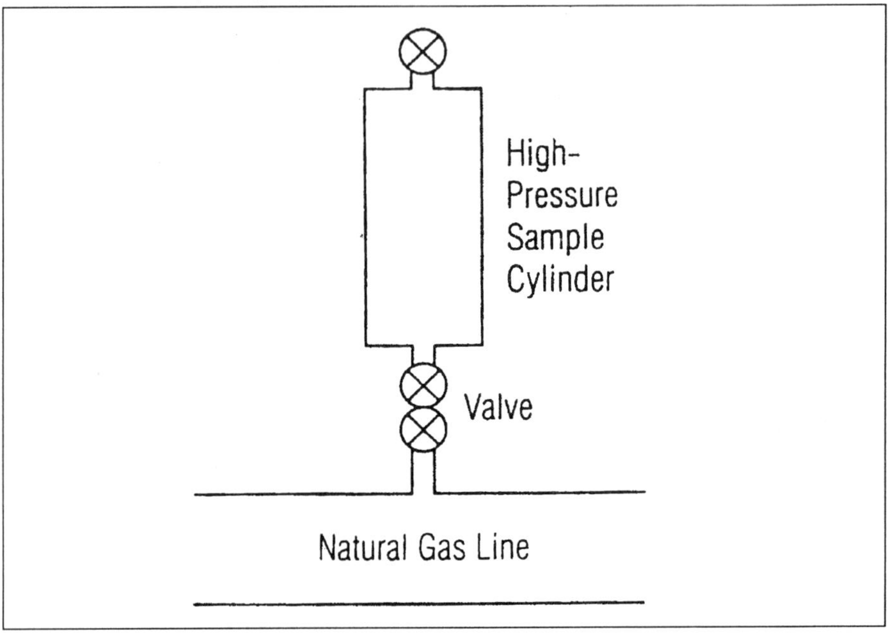

Radian assumed the total capture method was most representative, because no vapor or liquid was allowed to escape.

These three methods were used for sampling gas from a single source. The results demonstrated BTEX emissions based on GRI-DEHY [a predecessor model to GRI-CLYCalc v.2.0.] were:

Table 1–4 BTEX recovery

Method	BTEX
•TO-14	86 TPY
•GPA 2166-86	61 TPY
•Total Capture	90 TPY

The close correlation between TO-14 and the total capture methods strongly suggested the TO-14 was an acceptable method. Accordingly, Radian used the TO-14 method as a standard during later sampling.

Comparison analyses for BTEX for samples taken by TO-14 and GPA 2166-86 methods demonstrates the GPA method only captures a portion of the BTEX, Table 1–5.

Subsequent samples showed the GPA method produced varying ratios of individual components to total BTEX for the same gas. This phenomena suggests the GPA method gathered varying amounts of liquids [primarily water] from the pipe walls.

Table 1–5 Natural gas BTEX concentrations determined during preliminary field evaluation comparisons

	Concentration (ppmv)	
	modified EPA TO-14	**Common industry approach**
Benzene	218 ± 47	100 ± 28
Toluene	117 ± 34	75 ± 24
Ethylbenzene	6.7 ± 2.3	5.0 ± 3.5
Xylenes	32 ± 12	46 ± 26

Several features of the GPA 2166-86 may account for the differences between this method and TO-14 method:

- GPA method collects samples at gas line pressure
- GPA method includes a direct connection to the pipeline
  — collects liquids from pipe wall
  — provides no filter or trap to separate liquids from gas

Radian reduced the pipe wall effect by installing a venting manifold, Figure 1–3, and using an evacuated canister. The modified GPA method consisted of:

- evacuated canister
- manifold, Figure 1-3
- 30-min canister fill time
- final canister pressure equal to line pressure
- GC-MD analysis

Figure 1–3 Sampling manifold apparatus used in the detailed field comparison

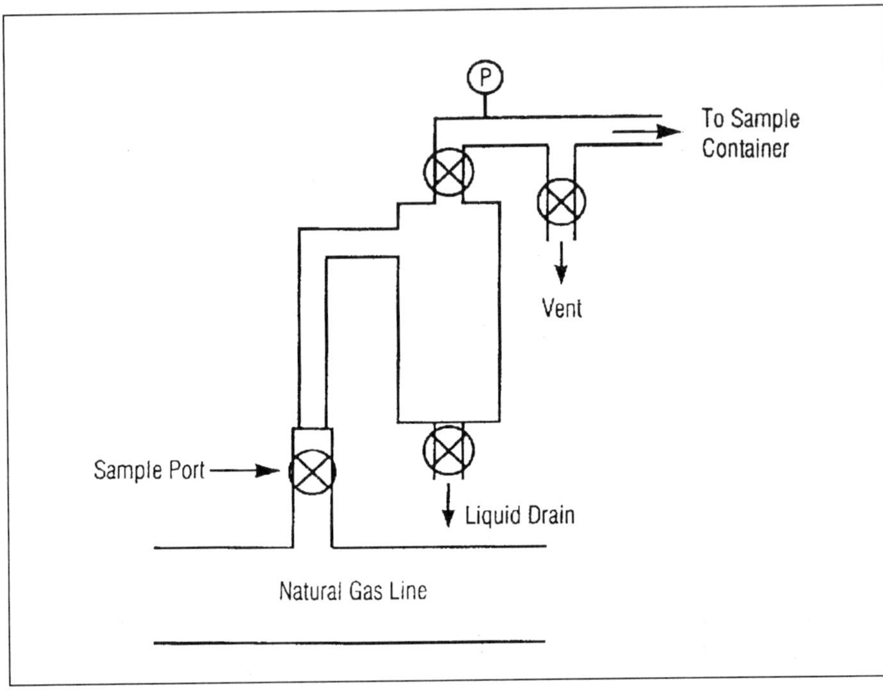

For convenience, the TO-14 method was modified to use the same manifold. The modified TO-14 method consisted of:

- evacuated canister
- manifold, Figure 1–3
- final canister pressure less than atmospheric pressure
- GC-TCD/FID analysis

Subsequent sampling at three plants in different states demonstrated the results from the two modified methods were the same, i.e., no statistical difference. *Based on these results, the standard GPA method [Figure 1–2] is not acceptable.*

The ability of gas to retain liquids is a strong function of temperature. Accordingly, when sampling in winter or under cold conditions, insulate the manifold and canister so the sample is taken at line temperature.

Maximum Achievable Control Technology

All major sources of hazardous air pollutants must apply MACT. By statute, MACT must represent the "most tightly controlled 12% of existing facilities," i.e., California controls.[42] Depending on the emitting source, several methods of control currently are available:

Dehydrator Emissions

1. Condense BTEX and return to the crude oil [some of the BTEX will be emitted from the tank, however, and a blow case or electricity is required to pump liquids to the tank]
2. Condense BTEX and burn in reboiler [requires blow case or electricity]
3. Condense BTEX and flare [requires agreement of the local air control board]
4. Condense BTEX and pump into pipeline

Tank Emissions

1. Vapor recovery on crude oil tanks will be required if the tanks or tanks and dehydrator are a major source of HAP. Two types of vapor recovery are available, motor-driven compression and produced water eductor system.
   - compressor type vapor recovery unit
     This unit is commercially available both for sweet and sour [H_2S] service. The unit consists of a compressor taking suction on the tank system, a heat exchanger to cool the compressed vapors, a pumping system to return the recovered condensed liquids to the crude oil or condensate tank, and a gas system to accept the recovered natural gas. A source of make-up gas is essential to avoid imploding the tank while running the tank liquids to the pipeline or truck. The major draw-back is the need for electric power for the vapor recovery unit [VRU].
   - eductor type vapor recovery unit
     This type of unit uses produced water flowing through a venturi to create a reduced pressure on the connected tanks. Typically, the produced water must be pumped through the eductor. This requires electric power. Once the tank vapors

[42] USC 4712 (d) (3) (A)

are induced into the system, they must be cooled via a heat exchanger and the liquid returned to the tank. Additionally, the recovered natural gas is compressed for injection into the available gas line.

Area Source – Hazardous Air Pollutants

Those facilities classed as area sources of HAP must apply generally available control technology [GACT]. The act provides no clue as to the interpretation of GACT. However, we expect such controls to be similar to MACT, because few controls exist which only partially reduce emissions of HAP.

Fortunately for E&P operators, the *Clean Air Act Amendments of 1990* exempt "oil and gas production wells (with its associated equipment) [from being listed] as an area source category..., except that the Administrator may establish an area source category for oil and gas production wells located in any metropolitan statistical area or consolidated metropolitan statistical area with a population in excess of 1 million...."[43]

Documentation[44] – Hazardous Air Pollutants

The CAA requires a "showing" of no increase in HAP emissions after the effective date of a state permit program. I believe this provision exempts facilities from major source classification if the operator reduces HAP emissions to below a major source level by the application of a non-MACT device *which increases oil or condensate recovery. The operator should document the recovery increase as the reason for installing the device.* Otherwise, there is a possibility the EPA or the state may allege the operator installed a non-MACT device as a means of circumventing MACT controls.

One innovative means of reducing BTEX emissions from glycol dehydrator vent lines is by use of the Texaco/Coastal recovery system. The system consists of convention air fins on the upper portion of the vent line and a wet glycol heat exchanger on the lower portion of the line. Typically, the system recovers sufficient BTEX to support the claim of "increased recovery." Installation prior to the advent of the

---

[43] 42 USC 7412 (n) (4) (B)

[44] 42 USC 7412 (g)

operating permit system program should eliminate the need for such a permit, if the emissions are reduced to below the HAP threshold.

OPERATING PERMITS — CRITERIA AND HAZARDOUS AIR POLLUTANTS

The EPA proposed the operating permit provision in the *Clean Air Act Amendments*[45] because of the success of the National Pollutionl Discharge Elimination System [NPDES] permit under the *Clean Water Act*. I expect the air permits will impose strict controls on major stationary sources and major sources, i.e., combustion and emitting equipment including fugitive emissions. CAA operating permits apply to stationary sources emitting:

- 100 TPY of any one criteria pollutant[46]
- 10 TPY of any one HAP[47]
- 25 TPY of any combination of HAP[48]

Permit Provisions

40 CFR §70.6 details content of operating permits.[49] Required parameters of the permit are:

- permit term may not exceed five years
- application submission within 12 months of state plan approval
- permit renewal application submition at least six months prior to expiration date of existing permit
- permit may be a general permit
- published permit notice
- opportunity for public participation
- general permit application, if permittee desires to operate under a general permit [may include standard exemptions]
- emissions monitoring, recordkeeping, and reporting
- permit fees

[45] Title V, 42 USC 7661
[46] 42 USC 7602
[47] 42 USC 7412 (A) (1)
[48] 42 USC 7412 (A) (1)
[49] 21 July 1992 at 57 FR 32250

Additionally, the permit may provide:

- operating permits for temporary sources
- emergency provisions allowing the state to issue a permit in less than the above stated time

Permit Fee

The CAA specifies the state is to collect fees sufficient to offset the operating permit program cost.[50] However, the state may increase the fee. Currently, a fee of $25 per TPY is contemplated,[51] i.e., $2,500 for a threshold major criteria pollutant emitting facility. States are not required to impose the fee on emitting sources in excess of 4,000 TPY.[52] However, at 40 CFR §70.9(b)(2)(ii), the EPA specifies the states *may* exclude sources emitting more than 4,000 TPY.

Permit Timing

Based on current information, the EPA or the states expect to grant operating permits for E&P operations during or before 1996. However, operators of stationary sources emitting 100 TPY or more and major sources of HAP must submit permit applications during November 1995. This requires the EPA to define MACT by that time. The EPA apparently is concerned about satisfying that requirement. Accordingly, at 58 FR 37778 on July 9, 1993, the EPA proposed operators of major sources submit permit applications proposing the MACT the EPA "would have" proposed.

NEW SOURCE PERFORMANCE STANDARDS

New Source Performance Standards [NSPS], codified as 40 CFR Part 60, affect E&P operations both onshore and offshore. NSPS have been adopted for turbines,[53] gas plants,[54] and sulfur plants.[55] Of these, the gas plant NSPS are most important to E&P operators.

[50] 42 USC 7661 (a) (b) (3)
[51] 40 CFR §70.9 (b) (2) (6)
[52] 42 USC 7661 (a) (b) (3) (B) (iii)
[53] 40 CFR Part 60 Subpart GG
[54] 40 CFR Part 60 subpart KKK
[55] 40 CFR Part 60 Subpart LLL

The EPA promulgated NSPS for gas plants as 40 CFR Part 60 Subpart KKK on June 24, 1985, at 50 FR 26124. The rule was effective January 20, 1984, the date of the proposal. The rule requires a gas plant constructed, reconstructed, or modified after January 20, 1984, to comply with Subpart KKK.

Subpart KKK requires all operators of affected facilities to comply with 40 CFR §60.482-1(a), (b) and (d) and §60.482-2 through §60.482-10. Certain exceptions apply as specified in §60.633.

General Standards[56]

- Demonstrate compliance within 180 days of start-up
- Demonstration relies on §60.485
- Equipment in vacuum services is exempted from provisions of §60.482-2 to 10

Pumps in Light Liquid Service[57]

- Pump monitored for leaks monthly using organic vapor analyzer [OVA]. [*Does not apply to non-fractionating gas plants having design capacities less than 10 MMSCFD.*]
- Visually inspect pump seals weekly for leaks.
- A leak is detected, a drip is observed, or the OVA reading is 10,000 or greater. [*Does not apply to non-fractionating gas plants having design capacities less than 10 MMSCFD.*]
- Repair leak within 15 days of discovery.
- First repair attempt must be made within five days of discovery.
- Dual seal pumps require special procedure.[58]
- Canned pumps are allowed.[59]

Compressors [60]

- Reciprocating compressors in wet gas [field gas] service are exempt from §60.482-3.

---

[56] 40 CFR §60-482-1

[57] 40 CFR§ 60.482.2

[58] 40 CFR §482-2 (c) (2)

[59] 40 CFR §482-2 (e)

[60] 40 CFR§ 60.482-3

- Compressors not in VOC [non-methane, non-ethane hydrocarbon ≤10 percent by weight] service are exempt from §60.482-3.
- Other compressors must be equipped with a barrier fluid seal system if:
  — barrier fluid system operates at a pressure greater than the compressor stuffing box pressure
  — barrier fluid system vents to a closed system
  — barrier fluid purges into process stream
- Barrier fluid system must be designed in accordance with rule §60.482-3.

Pressure Relief Devices in Gas/Vapor Service[61]

- Each pressure relief device must operate at an OVA reading of <500 ppm above background.
- Each pressure relief device, after a release, must be returned to operating condition within five days after release.
- Each pressure relief device may be attached to a closed vent system complying with §60.482-10(a) and (b).

Sampling Connection Systems[62]

- Each fluid sampling system must return purge fluid to system or closed vent.
- *In-situ* sampling systems are exempted.

Open-ended Valves and Lines[63]

- Each open-ended valve or line in VOC service must be plugged, capped, or blind flanged.
- Double block-and-bleed systems bleed valve may be open during operation but must remain closed at other times.

Valves in Gas/Vapor and Light Liquid Service[64]

- Monitor each valve monthly using an OVA instrument. [*Does not apply to non-fractionating gas plants having design capacities less than 10 MMSCFD.*]

---

[61] 40 CFR §60.482-4
[62] 40 CFR §60.482-5
[63] 40 CFR §60.482.6
[64] 40 CFR §60.482-7

- A leak is detected at an instrument reading of ≥10,000 ppm. [*Does not apply to non-fractionating gas plants having design capacities less than 10 MMSCFD.*]
- Any valve demonstrating no leak for two successive months may be monitored the first month of the next quarter, until a leak is detected.
- Leaks must be repaired within 15 days of discovery.
- First attempt repair must occur within five days of discovery.
- Canned valves are allowed.[65]

Pumps and Valves in Light Liquid Service; Flanges and Connectors in any Service[66]

- Pumps and valves in heavy liquid service, pressure relief devices in light liquid or heavy liquid service, or flanges and connectors in any service must be monitored using an OVA instrument within five days of suspecting a leak.
- A leak is detected at an instrument reading ≥10,000 ppm.
- Repair must occur within 15 days of discovery by the OVA.
- First attempt repair must occur within five days of discovery by the OVA.

Delay of Repair[67]

- Notwithstanding the 5- and 15-day repair requirement, repair may be delayed if repair is not possible without a process shutdown. Repair must start during the next shutdown.
- Repair of valves may be delayed, if the operator demonstrates emissions from purging the system will be greater than the leak if repair is delayed, but purged material must be captured.
- Repair of pump may be delayed for dual seal pumps.[68]
- Repairs may be delayed beyond the next shutdown for valves if:
  - — valve assembly replacement is required
  - — valve assembly supplies are depleted
  - — valve assembly supplies previously were on-hand
- Next shutdown must occur within six months of the last shutdown.

---

[65] 40 CFR §60.482-7 (h)

[66] 40 CFR §60.482-8

[67] 40 CFR §60.482-9

[68] 40 CFR §482-9 (d)

Closed Vent Systems and Control Devices[69]

- Closed vent systems and control devices must demonstrate 95 percent or more recovery efficiency.
- Combustion devices must demonstrate 95 percent destruction efficiency or provide at least 0.76-second residence time at 816° C [1,500° F].
- Flares must comply with 40 CFR §60.18.

RECORDKEEPING

Each device found leaking must be numbered using a tag affixed to the device.

The following records must be kept in the event of a leak:

- instrument and operator identification numbers and device number
- dates of leak detection and each repair attempt
- leak repair method attempted
- OVA reading observed after each repair attempt
- delayed repair – reasons for delaying repair if leak not repaired within 15 days [Each such delay reason must be signed by designated person.]
- expected repair date if repair delayed beyond 15 days
- process shut-down dates while device not repaired
- list of equipment designated "no detectable emissions," i.e., canned pumps or valves
- data demonstrating compressors are in wet gas service or non-VOC service

TITLE III — OUTER CONTINENTAL SHELF REQUIREMENTS

The *Clean Air Act Amendments of 1990* transfer control of Outer Continental Shelf [OCS] emissions for drilling and producing opera-

---

[69] 40 CFR §60.482-10

tions to the EPA from the Mineral Management Service [MMS]. The EPA promulgated OCS controls.[70] The rule mandates EPA responsibility for air pollution control of OCS operations located within 25 miles of the state seaward boundaries, except those located in the Gulf of Mexico west of 87.5 degrees longitude. Basically, the requirements within the 25-mile area are the same requirements applying to the corresponding onshore area [COA]. Jurisdiction for OCS operations outside the 25-mile line and within the 25-mile line outside of the *Submerged Lands Act* line west of 87.5 longitude remains with the MMS.[71]

The Memorandum of Understanding[72] defines the *Submerged Lands Act* line as "line of ordinary low water along that portion of the coast in direct contact with the open sea and line marking the seaward limit of inland waters [line of barrier islands]."

New sources subject to EPA control must comply with 40 CFR Part 55 on or after November 2, 1992. Existing sources which are stationary sources emitting 100 TPY or more and/or major sources, and located within 25 miles of state waters must comply on or before November 2, 1994. Additionally, 40 CFR Part 55 requires OCS operators, except in the Gulf of Mexico west of 87.5 longitude, to obtain operating permits for facilities emitting criteria pollutants at 100 TPY or more and/or HAP at 10 TPY [or 25 TPY for any combination of HAP] or more.

The act requires the Secretary of the Interior to complete, within three years [by November 1993] a study examining the impacts of OCS emissions on onshore Gulf Coast areas.[73] This could lead to controls imposed on Gulf of Mexico operations. The study has not been released.

---

[70] 40 CFR Part 55 on 4 Sept. 1992 at 57 FR 40792

[71] 33 CFR 250.45

[72] 59 FR 34102 on 1 July 1994

[73] 42 USC 7627 (b)

2 Clean Water Act

SUMMARY

The *Clean Water Act* controls discharges or potential discharges of contaminants into surface waters of the United States. E&P operations can involve one or all of the following permits or plans:

- National Pollutant Discharge Elimination System [NPDES] permit[1]
- storm water permits[2]
- spill prevention, control and countermeasure [SPCC] plan[3]

NPDES permits require a formal application and waiting period before granting. E&P facilities that have not experienced a reportable storm-related significant materials [oil] release to Waters of the U.S. during the past three years are exempt from applying for storm water permits. However, if a storm-related oil sheen or chemical release has entered Waters of the U.S., the operator is required in non-primacy states to submit to EPA a notice of intent [NOI] to comply with the existing general storm water permit.

---

[1] 40 CFR Part 122

[2] 40 CFR §§122.26 and 122.28

[3] 40 CFR § 112

SPCC plans apply to all facilities near Waters of the U.S., storing more than 660 gallons of hydrocarbons in a single tank or more than 1,320 gallons of hydrocarbons in more than one tank.

INTRODUCTION

The *Clean Water Act*[4] [CWA] is an outgrowth of the *Rivers and Harbors Act of 1899*,[5] an act designed to protect ships from damage due to logs floating freely in navigable waters. The *Federal Water Pollution Control Act*[6] was enacted on October 18, 1972, replacing the *Rivers and Harbors Act*. The last substantial revision to the *Federal Water Pollution Control Act* occurred in 1977[7] and resulted in the commonly known name of the *Clean Water Act*. The most recent revision, known as the *Oil Pollution Act of 1990*, is discussed in Chapter 3.

The CWA of 1977 includes three requirements impacting oil and gas operators:

- national pollutant discharge elimination system [NPDES] permit[8]
- storm water permit[9]
- spill prevention, control and countermeasure [SPCC] plan[10]

The goal of the CWA is to eliminate all discharges to Waters of the U.S.[11] The recent prohibition of any discharges from onshore E&P operations in states within EPA Region VI[12] [Figure 2–1] and discharges of drilling fluids and cuttings to coastal waters are the first steps toward reaching that goal.[13]

---

4 33 USC 1251
5 33 USC 441-451b
6 Public Law 92-500
7 PL 95-576
8 40 CFR Part 122
9 40 CFR Part 122
10 40 CFR Part 112
11 33 USC 1251 (a) (IO)
12 56 FR 7698, 25 Feb. 1991
13 58 FR 12454, 4 March 1993

Figure 2–1 EPA Regional Offices

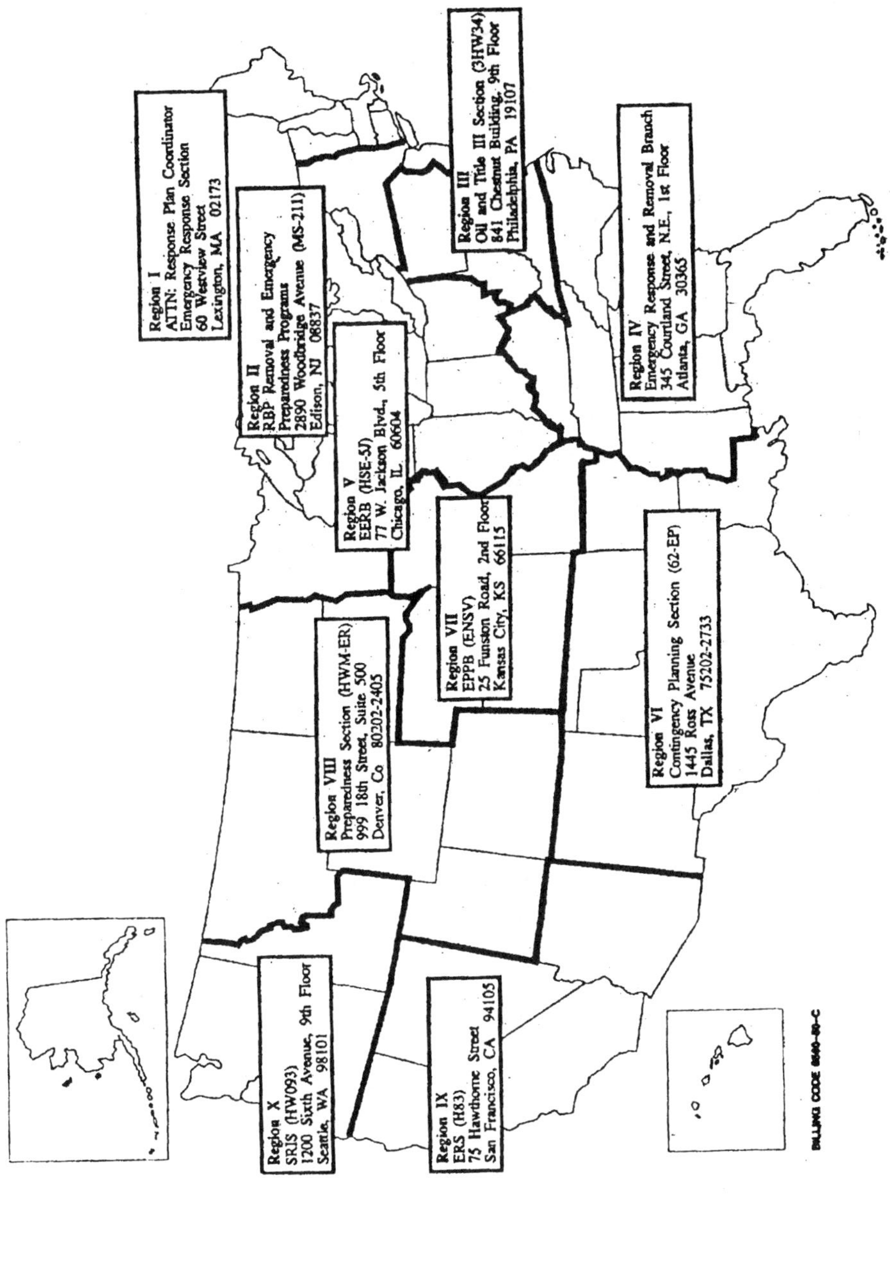

PERMITS

The CWA specifically authorizes point source discharge permits[14] into navigable waters. On the other hand, non-point source permits are not specifically mandated. However, the EPA uses the goals and policy provisions[15] as justification for the storm water permits.

National Pollutant Discharge Elimination System Permit

The CWA *prohibits* discharging of contaminated water into Waters of the U.S. without a current NPDES permit.[16]

Contaminated water includes waters containing oil, chemicals, total suspended solids [TSS], biological vectors or possessing high biochemical oxygen demand [BOD_5], chemical oxygen demand [COD], total organic carbon [TOC] or having a high or a low pH.[17]

Waters of the U.S. include oceans, lakes, rivers, streams, prairie potholes, playas, and dry stream beds which run during storms.[18]

Several types of applications for NPDES permits are allowed depending on the facility causing the discharge. Also, two types of permits are available for petroleum industry E&P facilities:

- general permits
- individual permits

Few onshore oil and gas extraction facilities require NPDES permits. The few requiring permits provide agricultural and wildlife water in the form of "fresh" produced water west of the 98th meridian, i.e., Rocky Mountain province.[19] On the other hand, most offshore facilities operate under general permits.

---

[14] 33 USC 1342

[15] 33 USC 1251

[16] 33 USC 1342

[17] 40 CFR §122.2

[18] 40 CFR §122.2

[19] 40 CFR §435.50

General NPDES Permits

The EPA invented general NPDES permits to lessen the regulatory burden on the agency.[20] Basically, the permit is issued by the EPA to the EPA. Accordingly, the EPA is the permit holder. Industrial dischargers meeting specific conditions may operate under the provisions of the permit by notifying the agency of their intent and agreement to comply with the general permit provisions. General permits are used onshore and offshore.

Effluent limitation guidelines for oil and gas facility discharges vary depending on geographical location.[21] The permit process is outlined in 40 CFR Part 122.

Onshore General Permits

Based on the effluent limitation guidelines,[22] EPA Region VI on February 25, 1991, at 56 FR 7698 issued general permits designated LAG 320000, NMG 320000, OKG 320000, and TXG 320000 for wells located onshore in Louisiana, New Mexico, Oklahoma, and Texas. These permits implemented the CWA goal of "no discharge."[23] Accordingly, these general permits prohibit all discharge from E&P facilities serving onshore wells [not in or over water]. The permits require separation of onshore well produced waters from coastal and offshore well produced waters. However, the general permits do not apply to agriculture and wildlife water use[24] or stripper[25] subcategories. Unlike the offshore permits, the onshore general permits require no notification or communication [no discharge monitoring report] with the EPA *unless provisions of the general permits are violated.*

To the best of my knowledge, the only other onshore general NPDES permit issued controls onshore discharges from hydrostatic testing of coal-bed methane pipelines and vessels issued by EPA Region VIII on December 13, 1991, at 56 FR 65085.

---

[20] 40 CFR §122.28

[21] 40 CFR Part 435

[22] 40 CFR Part 435

[23] 33 USC 1251 §101 (a) (1)

[24] 40 CFR Part 435 Subpart E

[25] 40 CFR Part 435 Subpart F

Offshore General Permits

EPA Regions IV and VI utilize general NPDES permits along the western Gulf of Mexico. Region VI issued a final NPDES general permit for the offshore subcategory of the oil and gas extraction point source category on November 19, 1992, at 57 FR 54642 and modified it on December 3, 1993, at 58 FR 63964. The modification became effective January 3, 1994.

This permit [GMG 290000] applies only to facilities located in waters of the Gulf Coast seaward of the outer boundary of the territorial seas of Louisiana and Texas.

The westerly boundary of the areas covered by the permit is defined as: Mobile, Viosca Knoll (north part), Destin Dome, Desoto Canyon, Lloyd and Henderson. The permit applies to all discharging stationary facilities seaward of a line three miles from the shore of Louisiana and Texas. Operators having state leases because of the Texas state mineral rights out to three leagues are required to request coverage [NOI] under the offshore general permit.

The NOI must be submitted at least 14 days before commencing of discharge. Certain lease blocks in or adjacent to MMS "no activity" areas or areas requiring live bottom surveys must include in the notice of intent the fact they are located in such areas and submit a notice of commencement of operations.[26] Further, these areas may require limiting maximum discharge rates for drilling fluids.

EPA Region VI modified general permit GMG 290000 on December 3, 1993, at 58 FR 63964. The modifications implemented revisions to 40 CFR Part 435, Subpart A dealing with:

- critical dilution for produced water
- monitoring requirements

Also, the modification requires compliance with the permit provisions within seven months of becoming aware of a non-compliance.

Permit GMG 290000 includes many provisions too numerous to discuss here. I recommend obtaining a copy of the general permit[27] before submitting a NOI. The permit expires November 19, 1997. Any

---

[26] GMG290000 Permit Part 1, §A.2

[27] 57 FR 54642 and 58 FR 63964

person submitting a NOI must notify the regional administrator by May 19, 1977, of intent to comply with the subsequent permit [content currently unknown].

Region VI proposed coastal water general NPDES permits for Louisiana [LAG 290000] and Texas [TXG 290000] on December 22, 1992, at 57 FR 60926. These permits would prohibit discharges of produced water and produced sand to coastal waters from facilities labeled:

- coastal subcategory[28]
- stripper subcategory[29]
- offshore subcategory[30]

On February 3, 1993, at 58 FR 6964 the EPA extended the comment period to March 15, 1993. As of May 1, 1995, the permits have not been issued.

Individual NPDES Permits

The vast majority of NPDES permits are issued under 40 CFR Part 122 as site and company specific permits, i.e., individual permits. Few individual permits are issued for E&P onshore and offshore operations. The facility operator makes application by completing specified forms[31] and submitting a description of the waste water treatment technology to be used and test data demonstrating the degree of contaminate removal provided by the technology. Typically, an NPDES permit requires about six months to a year to acquire depending on whether state or federal agencies grant the permit. Federal permitting requires more time.

Beneficial Use Discharge Permits

The most frequently granted individual NPDES permit authorizes the discharge of produced water of "good enough quality to be used for wildlife and livestock watering or other agricultural uses."[32]

---

[28] 40 CFR Part 435, Subpart D

[29] 40 CFR Part 435, Subpart F

[30] 40 CFR Part 435, Subpart A

[31] 40 CFR Part 122

[32] 40 CFR §435.51 (c)

Basically, a "beneficial use" permit is available to operators west of the 98th meridian, a line from Valley City, ND; Mitchel, SD; Neligh, NE; Superior, NE; Hutchinson, KS; Nash, OK; Duncan, OK; Bellevue, TX; Hico, TX; Seguin, TX; Alice, TX; and Rio Bravo, Mexico.

Cattle and wildlife can tolerate produced water having total dissolved solids [TDS] content ≤ 3500 mg/l. EPA regulations impose a 35 mg/l maximum oil and grease limitation[33] and restrict the discharge only to produced water. Typically, produced waters are discharged to dry washes or earthen tanks. In some cases, these produced waters enter navigable streams.

Typically, ranchers need "beneficial use water" and encourage EPA or state permitting of these discharges. However, some states have reduced the oil and grease concentration below the 35 mg/l allowed by the rule.

The EPA requires oil and grease determination on the basis of freon extraction and gravimetric analysis.[34] In certain areas of the Rocky Mountains, elemental sulfur is present in produced water. *If the sample is not filtered after extraction,* the sulfur is measured as oil and grease, making it very difficult to achieve the oil and grease permit provision.

Stripper Well Permits

EPA allows discharges of produced water into Waters of the U.S. from stripper wells, i.e., individual wells producing 10 or less bbl/day of oil.[35] Typically, this exemption applies in the Appalachian Region and Gulf Coast stripper wells. Stripper gas wells are not defined nor included within this exemption).[36]

Currently, stripper operators along the Gulf of Mexico coastal waters dispose of produced water into coastal waters under this exemption. See the discussion under Onshore General Permits for a proposed Region VI revision to this exemption.

Onshore Permit Guidelines

Best Practicable Control Technology [BPT] for onshore wells rep-

[33] 40 CFR §435.52 (a) (2)

[34] SW-846 Method 9070

[35] 40 CFR §435.60

[36] 40 CFR §435.61 (c)

resents "no discharge of waste water pollutants into navigable waters from sources associated with [producing operations]."[37] Accordingly, no discharge of oil and gas field waters is allowed from non-stripper wells into Waters of the U.S.

Permit Granting Agencies

The CWA[38] allows states "desiring to administer [the] permit program" ...to request "primacy" from the EPA. Thirty-nine states have been granted primacy.[39] However, many oil and gas producing states either have not applied or been granted primacy. These include as of May 1, 1995:

- Alaska
- Florida
- Idaho
- Louisiana
- New Mexico
- Oklahoma
- Texas

In these states, operators must apply to the EPA directly or to the EPA through the state agency. In some cases, two permits [state and EPA] will be required. Elsewhere, the state will process and approve or disapprove the permit application with EPA concurrance.

NPDES Permit Applicability

Any industrial point source discharging a pollutant into Waters of the U. S. on a continuing basis is required to hold a valid NPDES permit before commencing the discharge.[40]

Pollutant means "the man-made or man-induced alteration of the chemical, physical, biological, or radiological integrity of water."[41]

---

[37] 40 CFR §435.32 (a)

[38] 33 USC 1392 (b)

[39] 59 FR 5599, 7 Feb. 1994

[40] 33 USC 1342 (a) (1)

[41] 33 USC 1362 (19)

Pollutants are broadly defined. For the oil and gas extraction industry pollutants include:

- drilling fluids
- drill cuttings
- deck drainage or washdown water
- produced water
- produced sand
- well treatment fluids
- well completion and workover fluids
- sanitary wastes [treated or untreated]
- domestic wastes [gray water]
- miscellaneous fluids
  - — desalinization unit liquids
  - — blowout preventor fluids
  - — uncontaminated ballast water
  - — uncontaminated bilge water
  - — mud, cuttings, and cement at seafloor
  - — uncontaminated water (rain water)
  - — uncontaminated source water, diatomaceous earth filter media and excess cement slurry[42]
  - — filter backwash water

Point source refers to the discharge leaving a facility via "a discernible, confined, and discrete conveyance," i.e., a pipe, conduit, water course, or flowing stream.[43] This definition excludes ephemeral streams commencing on the facility property and disappearing before leaving the facility property.

Waters of the U.S. means all surface waters:

- oceans
- lakes
- streams
- wetlands
- intermittent streams when they flow [except ephemeral streams]
- prairie potholes
- ditches leading from the facility property which flow during rain.[44]

---

[42] 58 FR 63934

[43] 33 USC 1362 (14)

[44] 40 CFR §122.2

Discharges to publicly owned treatment works [POTW] are not subject to NPDES permitting.[45] However, such discharges are subject to pretreatment standards.[46]

A POTW permit is an acceptance on the part of the treatment works allowing the discharge to enter the treatment system. The degree of pre-treatment ranges from "nothing" to extensive "treatment." Typically, these conditions depend on the degree of contamination and the desire of the city to encourage industry.

NPDES Permit Provisions

NPDES permits are divided into sections:

- permit applicability and coverage conditions
- effluent limitations and monitoring requirements
- other discharge limitations
- other conditions
- standard conditions
- penalties for violations of permit conditions
- definitions

The provisions for general permits are not negotiable, because the Agency is the permittee. Nevertheless, modifications are possible as the result of comments submitted by the public.[47]

Conditions for individual permits can be negotiated but it takes time, patience, and supporting data.

The most important consideration regarding NPDES permits is not to accept a permit containing unachievable conditions, because the resulting penalties are too great. *Once a permit is accepted, virtually no means of revising the conditions downward are available!*

NPDES Effluent Limitations

The effluent limitations for the oil and gas extraction industry derive from 40 CFR Part 435 guidelines. These limitations represent best practicable control technology currently available as mandated

[45] 40 CFR §122.3 (c)

[46] 40 CFR Part 403

[47] 57 FR 54642, 19 Nov. 1992

by the CWA.[48] Section 1311 of the CWA mandates three technology based limitations:

- best practicable control technology currently available [BPT] [49]
- best conventional pollutant control technology [BCT] [50]
- best available technology economically achievable [BAT] [51]

Additionally, the EPA has instituted best professional judgment technology [BPJ] because BAT is not promulgated for all industries.

Best Practicable Control Technology

BPT was mandated for all industrial point source discharges effective July 1, 1977[52] and represents end-of-pipe treatment as opposed to in-plant controls. BPT also applies to "new dischargers" but not to "new sources."

New Discharges

A new discharger means a facility from which a discharge occurs on or after August 13, 1979, and which has not received a final NPDES permit for the discharge at that site.[53]

New Sources

A new source means a facility from which a discharge occurs or will occur (1) after promulgation of a new source performance standard *or* (2) after proposal of a new source performance standard if the standard is promulgated within 120 days of proposal.

Best Available Control Technology

BAT applies to existing and new discharges operating on or after July 1, 1988, and represents in-plant controls plus end-of-pipe treat-

---

[48] 33 USC 1311 (b) (1) (A)
[49] 33 USC 1311 (b) (1) (A)
[50] 33 USC 1311 (b) (2) (E)
[51] 33 USC 1311 (b) (2) (A)
[52] 33 USC 1311 (b) (1) (A)
[53] 40 CFR §122.2

ment.[54] Also, BAT applies to Toxic Pollutants, Appendix 2,[55] and Priority Pollutants, Appendix 3.[56]

Best Conventional Pollutant Control Technology

BCT applies primarily to publicly or privately owned treatment works [POTW]. However, BCT is incorporated in offshore permits.[57] BCT controlled pollutants include:

- biological oxygen demand [BOD_5]
- chemical oxygen demand [COD]
- total suspended solids [TSS]
- total organic carbon [TOC]
- fecal coliform
- ammonia as N
- pH
- oil and grease [O&G]

Best Professional Judgment Technology

The EPA defines BPJ as a case-by-case determination by the agency of BPT, BAT, or other treatment standards to achieve a particular pollution reduction.[58] The "judgment" refers to the permit writer's judgment based on EPA guidelines. Many existing permits are based on BPJ,[59] general permit for the Western Gulf of Mexico Outer Continental Shelf.

NPDES Permit Applications

Any industrial point source discharging contaminated waste liquids into Waters of the U.S. on a protracted basis is required to hold a valid NPDES permit before commencing the discharge.[60]

---

[54] 33 USC 1311 (b) (2)
[55] 40 CFR §401.15
[56] 40 CFR 423 Appendix A and 40 CFR §122.2
[57] 58 FR 53200, 14 Oct. 1993
[58] 57 FR 42572
[59] 57 FR 54642
[60] 33 USC 1342 (a) (1)

The application must be submitted at least 180 days prior to commencing the discharge.[61] Notwithstanding this requirement, the agency has no legal obligation to issue the permit in a timely fashion and the operator cannot discharge from the point source legally until the permit is granted. Permits are issued for a fixed term of not more than five years.[62] The permittee must apply for renewal of an existing permit at least 180 days before expiration.[63] If, through no fault of the permittee, a permit is not renewed on or before the expiration date, the existing permit continues in force, *if the permittee continues to comply with all provisions of the original permit.*[64]

NPDES Individual Permits

Applicants for individual permits must submit the following information on the forms provided by the director:[65]

1. the activity conducted requiring an NPDES permit
2. name, mailing address, and location of the facility for which application is made
3. Standard Industrial Classification [SIC] code describing the activity
4. operator's name, address, telephone number and ownership status
5. whether the facility is located on Indian lands
6. list of permits, construction approval received or applied for under:
   - *Resource Conservation and Recovery Act* — hazardous waste management
   - *Safe Drinking Water Act*
   - underground injection control
   - *Clean Air Act* — prevention of significant deterioration
   - *Clean Air Act* — non-attainment program
   - *Clean Air Act* — National Emission Standards for Hazardous Air Pollutants — preconstruction approval
   - *Marine Protection Research and Sanctuaries Act* — ocean dumping
   - *Clean Water Act* — dredge and fill[66]

[61] 40 CFR §122.21 (c) (1)
[62] 33 USC 1342 (b) (1) (B)
[63] 40 CFR §124.15
[64] 5 USC 558 (c)
[65] 40 CFR §122.21 (f)
[66] Section 404

7. topographic map extending one mile beyond the facility boundary.
8. brief description of the activity conducted

For existing facilities, applicants for individual permits must show:

1. outfall location; latitude and longitude to nearest 15 seconds
2. line drawing of water flow through facility
3. average flows and treatment
4. intermittent flows, if any
5. maximum production
6. compliance schedule, if any
7. effluent characteristics
   - BOD_5
   - COD
   - TOC
   - TSS
   - Ammonia [as N]
   - temperature [summer and winter]
   - pH

All data used in preparing the permit application must be retained for three years after submitting the application.[67] In the case of a corporation, the application must be signed by a corporate officer or his designated alternate. However, some leeway is permitted; see 40 CFR §122.22(a) (1) for details. For partnership, the application must be signed by a general partner.

NPDES General Permits

A general permit is granted by the EPA to the EPA. Typically, users of general permits submit to the EPA a NOI to comply with the permit. This NOI must be signed by the corporate officer or designated alternate. However, the onshore general NPDES permits prohibiting discharges for Louisiana, New Mexico, Oklahoma, and Texas[68] automatically apply to all E&P operations without need for a NOI.

---

[67] 40 CFR §122.2 (p)
[68] 56 FR 7698

Storm Water Permits

Storm water permitting was initiated by CWA Section 402(p)[69] added by the amendments of 1987. Traditionally, the EPA focused on industrial processes and municipal sewage discharges. However, with time, the EPA realized diffuse sources of storm water runoff caused receiving water pollution. Accordingly, on November 16, 1990, the EPA promulgated storm water regulations[70] imposing NPDES permit requirements on industrial facilities, large municipal separate storm sewer systems, and medium municipal separate sewer systems.

The following discussion of the storm water permitting regulations concentrates on industrial facilities, because this publication is aimed at industry.

Industrial Storm Water Permit Definitions

The EPA uses new terms in the storm water permit regulations:

- *Storm water discharge associated with industrial activities* means discharge from a conveyance used to collect and convey storm water directly related to manufacturing, processing, or raw material storage area.[71]
- *Significant materials* include raw materials, oil fuels, solvents, detergents, plastic pellets, finished metallic products, food processing associated with raw materials, hazardous substances,[72] any chemical reportable under SARA 313,[73] fertilizers, pesticides, waste, ashes, slag, sludge, or anything having the potential to be release into storm water.[74]

Storm Water Permit Applicability

The storm water regulations apply to any industrial activity discharging contaminated storm water [oil contaminated or containing

[69] 33 USC 1342 (p)
[70] 40 CFR §122.26 at 55 FR 47990
[71] 40 CFR §122.26 (b) (12)
[72] 40 CFR §302.4, Appendix A
[73] 40 CFR Part 372
[74] 40 CFR §122.26 (b) (12)

significant materials], via a conveyance [conduit, pipe, or man-made ditch] and enters or can enter Waters of the U.S.[75]

The definition of industrial activity includes the petroleum extraction industry subcategory. However, the EPA[76] exempts all existing or new discharges composed entirely of storm water from oil or gas exploration, processing or treatment [E&P] operations, or transmission facilities from submitting a storm water discharge permit application *unless*:

1. The facility has discharged storm water resulting in the release of a reportable quantity [RQ] of a hazardous substance[77] since November 16, 1987.
2. The facility has discharged storm water resulting in the discharge of a reportable quantity [causing a sheen on or sludge beneath the water][78] of hydrocarbons for which notification is or was required since November 16, 1987.
3. The facility storm water discharge contributes to a violation of a water quality standard.

Basically, the above exemption[79] allows clean E&P facilities to operate without storm water permits.

All existing industrial activities [except mining,[80] and E&P facilities not covered by the exemption] were required to submit a storm water permit application by October 1, 1992.

Storm Water Permit Availability

Non-exempt operators must apply for one of three permits:

- individual permit[81]
- group permit[82]
- general permit[83]

---

[75] 40 CFR §122.26
[76] 40 CFR §122.26 (6) (1) (iii)
[77] 40 CFR §302.4
[78] 40 CFR §§110.3 and 110.6
[79] 40 CFR §122.26 (6) (1) (iii)
[80] 40 CFR §122.26 (c) (1) (iv)
[81] 40 CFR §122.26 (c) (11)
[82] 40 CFR §122.26 (c) (2)
[83] 57 FR 41236 (Regions I, IV, VI, VIII, IX & X)

Individual Storm Water Permit

For existing sources composed entirely of storm water, submit EPA Form 2C [General Information Form] and Form 2F [Storm Water Discharge Form] including:

1. site map showing topography[84] including:
   - drainage conduits
   - paved areas, buildings, waste treatment areas, and material loading area
   - injection wells
   - springs and other bodies of water
2. estimate of area of impervious surfaces and total area drained by each outfall within one mile of the facility[85]
3. narrative description of significant materials:
   - treated onsite during the previous three years
   - stored or disposed in a manner allowing exposure to storm water
   - methods of treatment, storage, and disposal of these materials
   - materials management practices employed during the previous three years for these materials to "minimize" storm water contact with significant materials
4. location and type of existing structural and non-structural control measures to reduce storm water contamination
   - description of the treatment storm water receives and ultimate disposal of solid or fluid waste removed
   - certification that all outfalls containing storm water have been evaluated for presence of non-storm water; describe method used to make evaluations; provide date of evaluation and description of each outfall evaluated[86]
   - history of releases of toxic or hazardous pollutants occurring during previous three years[87]
   - analytical results from storm water samples taken from each outfall, including the following parameters that must be evaluated:[88]

---

[84] 40 CFR §122.26 (c) (i) (A)
[85] 40 CFR §122.26 (c) (1) (i) (B)
[86] 40 CFR §122.26 (c) (1) (i) (C)
[87] 40 CFR §122.26 (c) (1) (i) (D)
[88] 40 CFR §122.26 (c) (1) (i) (E)

a. oil and grease, pH, BOD_5, COD, TSS, total phosphate, total Kjeldahl nitrogen, nitrates, and nitrite nitrogen
b. flow rate [estimated or measured], total flow per storm event sampled, and method of measurement or estimate
c. date and duration [in hours] of storm event sampled, rain fall amount [measured or estimated in inches] during event sampled, and previous storm generating more than 0.1" rainfall

For existing sources composed of storm water and non-storm water, submit EPA Forms 1, 2C, and 2F. Also, comply with all of the above requirements.[89]

For new sources or discharges composed of storm water and non-storm water, submit EPA Forms 1, 2D and 2F.[90]

Group Storm Water Permit

If a group of similar dischargers desire, they may submit a group application with two parts:

1. Part 1[91] includes:
   - summary of each participant's industrial activity
   - explanation of why the participants are sufficiently similar to use a group application
   - list of significant materials not under roof and description of material management practices
   - list of discharges to be sampled and analyzed [This list must include at least 10% of the group outfalls or as required by Part 2 below.]
2. Part 2[92] — submit sample and analytical data for at least 10 percent of the discharges [a maximum of data from 100 dischargers] and at least two dischargers per each precipitation zone,[93] if at least 10 participants operate in that precipitation zone, or at least one discharger per precepitation zone, if less than 10 participants operate in that precipitation zone.

---

[89] 40 CFR §122.26 (c) (1) (i) (F)
[90] 40 CFR §122.26 (c) (1) (i) (G)
[91] 40 CFR §122.26 (c) (2) (i)
[92] 40 CFR §122.26 (c) (2) (ii)
[93] 55 FR 48073, 16 Nov. 1990

The EPA gave notice on March 21, 1991, at 56 FR 12098 that group permit Part 1 applications for existing sources must be received by September 30,1991. The EPA has 60 days during which to approve or disapprove group permits. Part 2 application must be received by May 18, 1992, or within 12 months after approval of Part 1, whichever comes first.

For new discharges, the EPA requires a permit application be submitted 180 days before commencing the discharge. The group permit concept has not been embraced by the regulated industrial community, primarily because of the extensive data submission provisions and the necessary interaction between group participants. Nevertheless, the EPA proposed a draft general storm water group permit on November 19, 1993.[94] This permit exempted oil and gas operators from submitting a NOI if no reportable storm water contaminated release had occurred during the past three years. Refer to the General Permit Application for details. This draft permit applied to industrial operators submitting group permit applications in Regions, I, II, III, IV, V, VI, VIII, IX, and X.[95] As of May 1, 1995, no final permit has been issued.

General Storm Water Permit

The EPA issued a general storm water permit for industrial activities on September 9, 1992, at 57 FR 41236. The general permit applied to all industrial groups including oil and gas extraction industry operating in Regions I, IV, V, VI, VIII, IX, and X. However, an oil and gas extraction facility is exempt, if the facility has not nor was required to submit a storm water-related release report of oil or hazardous substance within the past three years preceding September 9, 1992. This exemption[96] applies to:

- crude petroleum and natural gas[97]
- natural gas liquids[98]
- drilling oil and gas wells[99]

---

[94] 58 FR 61146
[95] 58 FR 61146
[96] 58 FR 61196
[97] SIC Code 1311
[98] SIC Code 1321
[99] SIC Code 1381

- oil and gas field exploration services[100]
- oil and gas field services, not elsewhere classified[101]

If a storm water release report was submitted or required, the NOI to comply with the general permit must be submitted on or before October 1, 1992. If a reportable release occurs after October 1, 1992, the operator must submit a NOI within 14 days of the incident.

All notices of intent, notices to terminate, storm water pollution prevention plans, reports, or information regarding the permit must be signed:

- for a corporation — by the president, secretary, treasurer, or vice president of the corporation or other person in charge of a facility employing more than 250 persons or having a gross annual sales or expenditures exceeding $25 million, if the delegation has been given in accordance with corporate procedures
- for partnership or sole proprietorship — by a general partner or the proprietor, respectively [102]

Based on a review of the general permit provisions,[103] I believe an operator should use the general permit because of the ease of application and the reduced requirements, if not exempted.

Each industry permittee except E&P is required to sample storm water twice per year and submit the analytical results to the permit granting agency. E&P permittees are exempt from this provision.[104]

Storm Water Pollution Prevention Plan

The storm water rule requires preparation of a pollution prevention plan [PPP]. The general permit requires preparation of a PPP within six months of October 1, 1992 [April 1, 1993] or the date of submission of the NOI, whichever is later.[105]

The plan must include a description of the facility, description and quantity of significant materials [oil and hazardous substances], mea-

[100] SIC Code 1382
[101] SIC Code 1389
[102] 57 FR 41236
[103] 57 FR 41236
[104] 57 FR 41236
[105] 57 FR 41236

sures to prevent releases of significant materials and a contingency plan for control and countermeasure activities.[106] In effect, the PPP is a glorified SPCC plan.[107]

Unlike the SPCC plan, the PPP need not be signed by a professional engineer. Instead, the PPP must be signed by a company officer having the authority to sign the NOI. [See General Permits above]

Further, the PPP must require annual compliance audits of the facility. The facility must be in compliance with the PPP and all of its provisions. The compliance audit must be signed by the same person submitting the NOI.

This signatory requirement forces the delegated person to assume responsibility for compliance. Additionally, it assures imposition of criminal penalties if the audit results are falsified or ignored.

SPILL PREVENTION, CONTROL, AND COUNTERMEASURE PLANS

The CWA[108] mandates preparation of spill prevention, control, and countermeasure [SPCC] plans.

The EPA promulgated SPCC regulations 40 CFR Part 112 on December 11, 1973, at 38 FR 34165. These regulations apply to any non-transportation-related onshore facility and facilities located landward of the coastal water line[109] engaged in operations that could result in the release of hydrocarbons into or upon the Waters of the U.S.

The regulations do not apply to the following:

1. Facilities that due to location reasonably could not be expected to discharge hydrocarbons into or upon Waters of the U.S. This determination must be made on the basis of geographical and locational aspects of the facility, i.e., proximity to waters, land contour, etc. Man-made containment features are excluded from consideration.
2. Equipment or operators of vessels or transportation-related onshore or offshore facilities that are subject to the authority and control of the Department of Transportation, i.e., drums.

[106] 57 FR 41236, Part IV, c.2.

[107] 40 CFR Part 112

[108] 33 USC 1251

[109] 59 FR 34102

3. Facilities having underground storage tanks with a capacity of 42,000 gal [1,000 bbl] *or* less or above ground storage tanks of 1,320 gal or less, provided no single tank has a storage capacity in excess of 660 gal.

Accordingly, the SPCC plan regulations apply to all E&P tank batteries or single storage tanks containing more than 660 gal crude oil, condensate, diesel fuel, lubricating oil, gasoline, solvent, or kerosene onshore and in most state waters.

SPCC Plan Requirements

The SPCC plan regulations[110] require any operator who, because of location and quantity of stored hydrocarbons, can be expected to release hydrocarbons in harmful quantities [sheen on water] into Waters of the U.S., to prepare and implement an SPCC plan within six months of beginning operations.[111]

In the case of drilling rigs, workover rigs, or mobile offshore rigs, the SPCC plan may be rig specific.[112] SPCC plans for stationary sources must be site specific. All SPCC plans for stationary sources must be reviewed and certified by a professional engineer [P.E.]. The plan must be retained at a site normally attended at least eight hours per day.[113] The SPCC plan only needs to be prepared for onshore facilities and those in or over state waters.

SPCC Plan

The SPCC plan should include the following:[114]

- tank number and sizes
- total storage volume
- direction of oil flow, if released
- rate of oil flow, if released
- size of dike, if so equipped
- diversionary structures, if so equipped

[110] 40 CFR Part 112
[111] 40 CFR 112.3 (b)
[112] 40 CFR §112.3 (c)
[113] 40 CFR §112.3 (f)
[114] 40 CFR §112.7

- if diversionary structure not practicable, provide a contingency plan
- spill remediation equipment, if available
- signed management commitment to provide necessary personnel, equipment and funds to remediate the oil spill.

The American Petroleum Institute [API] developed plan forms [D-16] including these items in a convenient format. The forms are available from the API's Dallas Office covering three separate operations, (1) drilling, (2) producing, and (3) gas plants.

Regulation 40 CFR §112.7(d) requires that a contingency plan be included in the SPCC plan, if all of the above conditions cannot be satisfied. Typically, diking of flowlines is not practical. Accordingly, a contingency plan is required for virtually all oil and gas extraction facility SPCC plans. Additionally, 40 CFR §112.7(d) requires a demonstration of such impracticality. A statement regarding the difficulty of diking flowlines provides the needed demonstration. Guidelines for contingency plan preparation are available in 40 CFR Part 109.

SPCC Contingency Plan[115]

- identify waters and uses
- establish notification procedure
- define authorities, responsibilities, and duties of persons involved
- list of names, telephone numbers, and addresses of responsible persons and agencies
- establish a communications system
- inventory of available equipment, if available
- define actions to be taken after release discovery

No SPCC plan is required for OCS activities or state waters. However, MMS requires a contingency plan[116] for OCS and some state waters.

[115] 40 CFR 109.51
[116] 30 CFR §250.42

SPCC Plan Submission

SPCC plans need not be submitted to the EPA or the state for approval unless the operator discharges 1,000 gal or more of oil into Waters of the U.S. during a single spill event or discharges two harmful quantity spill events within a 12 month period.[117] *Harmful spill event* is defined as forming a sheen on the water surface or sludge under the water surface.[118]

SPCC Plan Training

Typically, SPCC plans contain a requirement to train personnel in the content and use of the plan.[119] This training generally consists of a review of the plan and employee responsibilities in the event of a release.

MMS OIL SPILL CONTINGENCY PLAN

MMS requires submission of the contingency plan for approval prior to or with the exploration plan or development and production plan[120] and annually thereafter. The *Oil Pollution Act of 1990* requires submission of contingency plan for state water facilities in direct contact with the ocean as a substitute for a response plan until February 18, 1995. Refer to Chapter 3 for details.

MMS Plan Training

MMS requires annual hands-on training in deploying and operating the oil spill response equipment.[121] Additionally, the supervisory personnel must be trained in directing deployment and use of the response equipment. The operator must maintain course-completion certificates or attendance records.[122]

[117] 40 CFR §112.4 (a)
[118] 40 CFR §110.3
[119] 40 CFR §112.7 (e) (10)
[120] 30 CFR §250.42
[121] 30 CFR 250.43
[122] 30 CFR §254 at 58 FR 7489

Drills to familiarize personnel with the response equipment must be held when the equipment is placed in service initially and annually thereafter. The oil spill response team designated in the oil spill contingency plan must participate in the drills. The drills must simulate conditions in the area of anticipated release and must include deployment of the equipment. The drill schedule must be submitted to the MMS regional supervisor sufficiently in advance to allow MMS personnel to witness the drill [typically two weeks in advance].

Records of drill conditions, results, and names of participants must be recorded and maintained for two years. Refer to the National Preparedness for Response Exercise Program [PREP] Guidelines in Chapter 3 for details.

3 Oil Pollution Act Of 1990

SUMMARY

Congress passed the *Oil Pollution Act of 1990*[1] [OPA-90] as a means of reducing oil spills by increasing surveillance and preparedness and requiring spill response plans for all transportation- and non-transportation-related onshore and offshore facilities storing, processing, or transporting oil above certain threshold quantities near Waters of the U.S.

The president assigned implementation of OPA-90 to four Agencies:[2]

- Environmental Protection Agency [EPA] — Onshore non-transportation-related facilities [tank batteries]
- Research and Special Programs Administration [RSPA] — Onshore transportation-related facilities [pipelines], Department of Transportation [DOT]
- U.S. Coast Guard [USCG] — Offshore transportation-related facilities [ship and loading facilities], Department of the Interior [DOI]
- Mineral Management Service [MMS] — Offshore non-transportation related facilities [platforms] [DOI]

---

1 PL 101-380

2 E.O. 12777

This division of jurisdictions resulted in regulatory problems within the agencies. Accordingly, on July 1, 1994, the EPA published a Memorandum of Understanding [MOU] returning the jurisdiction to a more traditional arrangement.

Typically, onshore E&P facilities are exempted from OPA-90, because the triggering threshold quantity is too high [23,810 bbl]. On the other hand, offshore platforms invariably must prepare and submit a response plan. Only a few E&P facilities are subject to RSPA and/or USCG rules because of pipelines and marine loading operations. Compliance with OPA-90 is important, because the penalty for contravening OPA-90 includes prohibition on producing, processing, or transporting oil at or from the facility.

INTRODUCTION

The publicity surrounding Exxon Valdez tanker spill in Prince William Sound in March 1989 allowed Congress to enact the OPA-90 as an amendment to the *Clean Water Act*.[3] The major thrust of the act is aimed at tanker pollution. However, onshore, coastal, Outer Continental Shelf [OCS], and liquid pipeline operations are impacted by OPA-90. The act requires preparation of a detailed response plan for existing facilities by February 18, 1993, and prohibits any operations controlled by the regulations, if the response plan is not approved by August 18, 1993.

AGENCIES INVOLVED

The president vested implementation of OPA-90 in four Agencies:

- EPA
- RSPA [DOT]
- USCG [DOI]
- MMM [DOI]

The Agency assignments are on the basis of transportation or non-transportation related facilities. Accordingly, depending on the cir-

[3] 33 USC 2701

cumstances, more than one agency can be involved at one facility. When this occurs, the facility is called a "complex."[4]

EPA RESPONSE PLAN

Jurisdictional Responsibility

On July 1, 1994, the EPA promulgated at 59 FR 34070, the Oil Pollution Prevention Rule, 40 CFR Part 112, requiring preparation of a response plan for non-transportation-related onshore facilities by February 18, 1993. EPA proposed this rule on February 17, 1993, at 58 FR 8824 as a guide to compliance with OPA-90 "on or before February 18, 1993." EPA at 40 CFR §112.2(c) defines "onshore" as meaning "any facility located in, on or under any land within the United States, other than submerged lands..."

A MOU signed February 3, 1994,[5] gives the EPA jurisdiction over all non-transportation-related fixed [offshore] facilities located landward of the coast line.[6] Accordingly, EPA has jurisdiction over onshore and offshore non-transportation related facilities landward of the Submerged Land Act Line. The memorandum defines the coastline per the *Submerged Land Act*[7] [SLA] as the:

- Line of ordinary low water along that portion of the coast in direct contact with the open sea
- Line marking the seaward limit of inland waters, meaning the line of ordinary low water along the seaward side of the barrier island

In Texas, the SLA coast line extends from the Mexican border northeast along the seaward edge of South Padre Island, Padre Island, Matagorda Island, Matagorda Peninsula, Follets Island, Galveston Island, Bolivar Peninsula, and along the coast line to the Louisiana border,

In Louisiana, the SLA coast line extends along the Sabine River eastward along the coast to Marsh Island and thence along the seaward edge of Marsh Island, Pointe Au Fer Island, Isle Dernures, Casse

[4] 59 FR 34072
[5] 59 FR 34102
[6] 59 FR 34071
[7] 43 USC 1301(C)

Tete Island, along the coast of Lafourche Parish, then along the seaward edge of Grand Isle and the barrier Island of Barataria Bay, thence along the south and north coasts of Plaquemines Parish to Breton Island and along the seaward edge of Brenton and Chandeleur Island.

Further, the MOU delegates to the EPA responsibility for non-transportation-related offshore facilities located landward of the coast line, i.e., in the Gulf of Mexico, all state waters shoreward of the barrier island. The EPA has not proposed or promulgated non-transportation-related offshore facility regulations. The agency plans to prepare such a rule at a later date and suggests using the onshore rule as guidance.[8]

Response Plan Preparation Criteria – Onshore Facilities

The threshold criteria for requiring preparation and submission of a response plan depend upon the facility. The EPA established the criteria for onshore non-transportation-related facilities, e.g., refineries, tank batteries, etc. A facility may perform several functions, i.e., store and ship oil. Accordingly, one facility may be controlled by more than one agency. When this occurs, the facility is called a "complex." Being a complex triggers preparation of a response plan at the lowest threshold level of the cognizant agency. A complex may prepare a single response plan. However, all of the plan elements required by each agency must be included in separate sections.[9]

For EPA, the threshold requiring a response plan [Figures 3–1 and 3–2] includes:

- over water transfer of oil from or to a vessel or barge from an onshore facility or facility located shoreward of the coast line having storage capacity ≥ 1,000 bbls[10] *or*

[8] 59 FR 34072

[9] 59 FR 34122

[10] 40 CFR §112.20 (f) (1) (i)

Figure 3–1 OPA-90 response plan flow chart

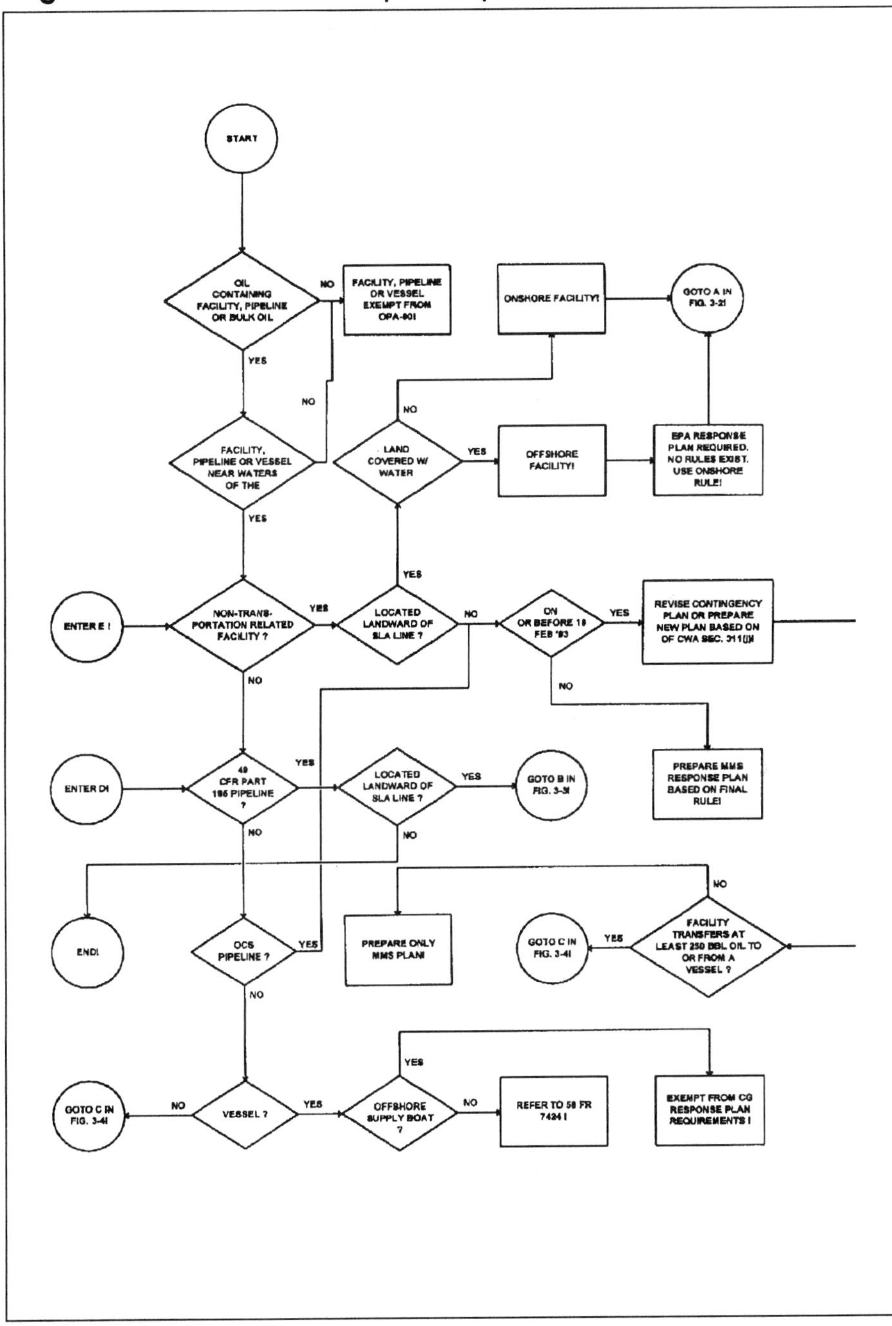

Figure 3–2 EPA response plan flow chart

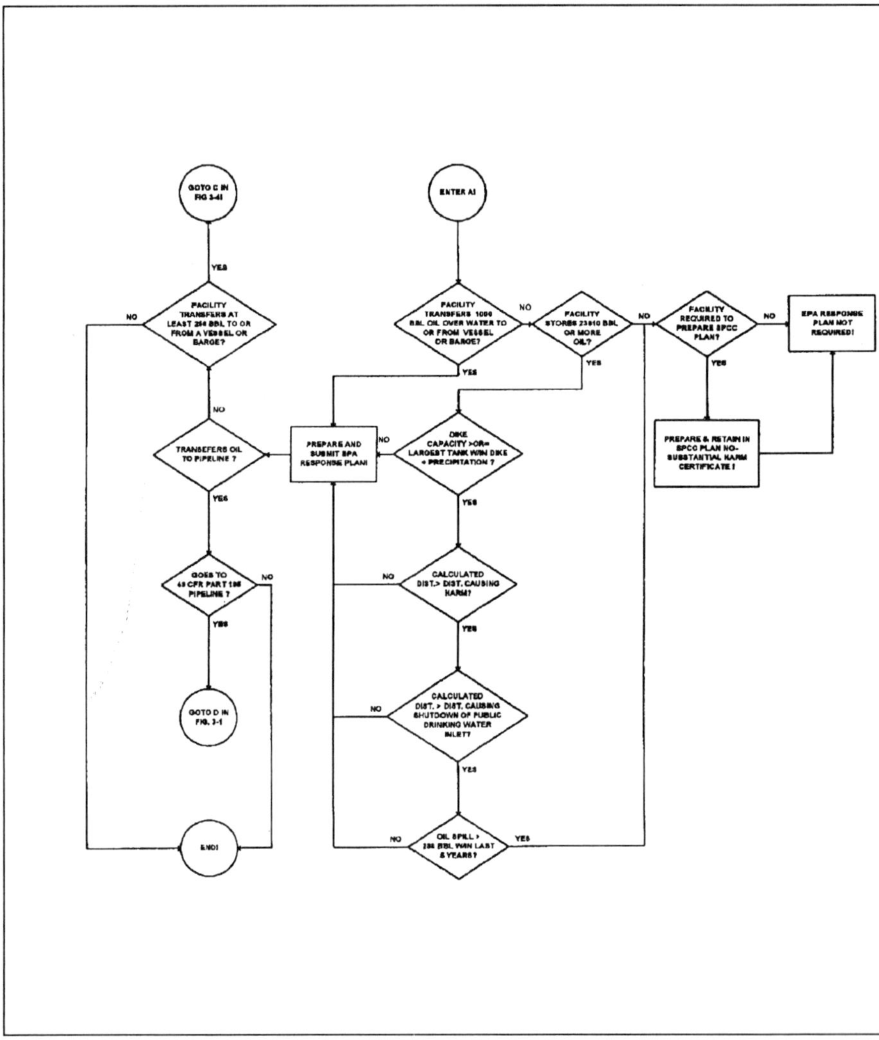

- Onshore facility having a storage capacity ≥ 23,810 bbl *and*
  — facility does not have secondary containment sufficient to contain volume of the largest tank within the containment plus precipitation [I suggest 25-year storm precipitation] *or*

—facility is located at a calculated distance[11] from the facility that would cause injury to fish, wildlife, and/or sensitive environments *or*
— facility is located at a calculated distance[12] from the facility that would cause shut-down of a public drinking water intake *or*
— facility has experienced a ≥ 238 bbl oil spill within the last five years

If any one of the above threshold criteria is satisfied, a response plan must be prepared and submitted to the EPA, Figure 3–2. Figures 3–1 and 3–2 provide a logical method for determining if a facility requires an EPA response plan. Enter Figure 3–1 at "start" and answer each question "yes" or "no" until a determination is made.

Typically, no U.S. onshore E&P facility [tank battery] contains 23,810 bbl of storage capacity. Accordingly, all onshore E&P tank batteries are exempted from OPA-90 control. On the other hand, many offshore facilities have a storage capacity of 1,000 bbl and transfer crude oil, condensate, or diesel fuel over water from or to a ship or barge. These facilities must prepare, implement, and submit a response plan to the EPA.

EPA Response Plan Submission

The rule[13] became effective August 30, 1994 and requires:[14]

- existing facilities in operation on February 18, 1993, having submitted a response plan on or before February 18, 1993, revise the plan to conform to the rule on or before February 18, 1995
- existing facilities in operation on February 18, 1993, failing to submit to the regional office a response plan on or before February 18, 1993, submit a plan on or before August 30, 1994

[11] Attachment C-III at 59 FR 34105
[12] Attachment C-III at 59 FR 34105
[13] 40 CFR Part 112
[14] 40 CFR §112.20 (a)

- facilities commencing operations after February 18, 1993, but prior to August 30, 1994, must submit to the regional office a plan and completed response plan cover sheet, Form 3–1,[15] or before August 30, 1994
- facilities commencing operations after August 30, 1994, must submit to the regional office a plan *and* completed response plan cover sheet, Form 3–1, on or before commencing operations [Revisions to plan caused by "startup" must be submitted immediately after completion of a 60-day operational trial period.]
- facilities requiring response plan preparation or revision after August 30, 1994, caused by *planned*:
  —Design changes
  —Construction changes
  —Operating changes
  —Maintenance changes
  must submit to the regional office a plan and completed response plan cover sheet, Form 3–1, before commencing operation of the portion of the facility subject to the change [Revisions to the plan caused by "startup" must be submitted immediately after completion of a 60-day operational trial period.]
- facilities requiring response plan preparation or revision after August 30, 1994, caused by an *unplanned* event or change must submit a plan and completed response plan cover sheet, Form 3–1, within six months after the unplanned event or change

[15] 59 FR 34136

Form 3–1 Response plan cover sheet

This cover sheet will provide EPA with basic information concerning the facility. It must accompany a submitted facility response plan. Explanations and detailed instructions can be found in Appendix F.[16] Please type or write legibly in blue or black ink. Public reporting burden for the collection of this information is estimated to vary from 1 hour to 270 hours per response in the first year, with an average of 5 hours per response. This estimate includes time for reviewing instructions, searching existing data sources, gathering the data needed, and completing and reviewing the collection of information. Send comments regarding the burden estimate of this information, including suggestions for reducing this burden to: Chief, Information Policy Branch, PM-223, U.S. Environmental Protection Agency, 401 M St., SW., Washington, D.C. 20460; and to the Office of Information and Regulatory Affairs, Office of Management and Budget, Washington D.C. 20503.

General Information

Owner/Operator of Facility: ____________________

Facility Name: ____________________

Facility Address (street address or route): ____________________

City, ____________________ State, ____________ Zip Code: ________

Facility Phone No.: ____________________

Latitude (Degrees: North): ____________ degrees, minutes, seconds

Dun & Bradstreet Number:[17] ____________________

Largest Aboveground Oil Storage Tank Capacity (Gallons): ____________

Number of Aboveground Oil Storage Tanks: ____________

Longitude (Degrees: West): ____________ degrees, minutes, seconds

Standard Industrial Classification (SIC) Code: ____________

Maximum Oil Storage Capacity (Gallons): ____________

Worst Case Oil Discharge Amount (Gallons): ____________

Facility Distance to Navigable Water. Mark the appropriate line.

0-1/4 mile ______ 1/4-1/2 mile ______ 1/2-1 mile ______ >1 mile ______

[16] 59 FR 34122

[17] These numbers may be obtained from public library resources.

Facility owners required to prepare a Spill Prevention Control and Countermeasure [SPCC] plan[18] and not satisfying any one of the threshold criteria, must prepare and sign a Certificate of Non-Substantial Harm and retain the Certificate in the SPCC plan[19] using Form 3–2.[20]

Form 3–2 Applicability of substantial harm criteria

Does the facility transfer oil over-water[21] to or from vessels and does the facility have a total oil storage capacity greater than or equal to 42,000 gallons?
Yes ________ No ________

Does the facility have a total oil storage capacity greater than or equal to 1 million gallons and, within any storage area, does the facility lack secondary containment[21] that is sufficiently large to contain the capacity of the largest aboveground oil storage tank plus sufficient freeboard to allow for precipitation? Yes ________ No ________

Does the facility have a total oil storage capacity greater than or equal to 1 million gallons and is the facility located at a distance[21] [as calculated using the appropriate formula in Appendix C-III (59 FR 34105) or a comparable formula] such that a discharge from the facility could cause injury to fish and wildlife and sensitive environments?[22]
Yes ________ No ________

Does the facility have a total oil storage capacity greater than or equal to 1 million gallons and is the facility located at a distance[21] (as calculated using the appropriate formula in Appendix C or a comparable formula) such that a discharge from the facility would shut down a public drinking water intake?[21] Yes ________ No ________

Does the facility have a total oil storage capacity greater than or equal to 1 million gallons and has the facility experienced a reportable oil spill[21] in an amount greater than or equal to 10,000 gallons within the last 5 years? Yes ________ No ________

Certification
I certify under penalty of law that I have personally examined and am familiar with the information submitted in this document, and that based on my inquiry of those individuals responsible for obtaining information, I believe that the submitted information is true, accurate, and complete.
Signature: ________
Name (Please type or print): ________
Title: ________
Date: ________

---

[18] 40 CFR Part 112

[19] 40 CFR §112.20 (e)

[20] 59 FR 34105

[21] Explanations of the above-referenced terms can be found in Appendix C to this part. If a comparable formula to the ones contained in Attachment C-III is used to establish the appropriate distance to fish and wildlife and sensitive environments or public drinking water intakes, documentation of the reliability and analytical soundness of the formula must be attached to this form.

[22] For further description of fish and wildlife and sensitive environments, see Appendices I, II, and III to DOC/NOAA's "Guidance for Facility and Vessel Response Plans: Fish and Wildlife and sensitive Environments" (see Appendix E to this part, section 10, for availability) and the applicable ACP.

EPA Response Plan Content

Each response plan submitted to the EPA must follow the model plan format[23] unless a state or other equivalent federal response plan is submitted. If another plan is substituted, the plan must include an "emergency response action plan"[24] and cross-reference section including:[25]

- facility information
- information about emergency response
- hazard evaluation
- response planning levels
- discharge detection systems
- plan implementation
- self-inspection
- diagrams
- security systems

Model EPA Plan Format

The EPA requests use of the model plan format included in 40 CFR Part 112, Appendix F at 59 FR 34122. The model plan includes:

Emergency Response Action Plan

- a compilation and repetition of:
  — qualified individual information
  — emergency notification phone list
  — spill response notification form
  — response equipment list and location
  — response equipment testing and deployment
  — evacuation plan
  — immediate actions
  — facility diagram

[23] 59 FR 34122
[24] 59 FR 34100
[25] 59 FR 34100-34101

Facility Information

- use facility information form,[26] Form 3–3

Form 3–3 Facility information form

Facility Name: ______
Location (Street Address) ______
City: ______ State: ______ Zip: ______
County: ______ Phone Number: ______
Latitude: ______ Degrees ______ Minutes ______ Seconds ______
Longitude: ______ Degrees ______ Minutes ______ Seconds ______
Wellhead Protection Area: ______
Owner: ______
Owner Location (Street Address): ______
(if different from Facility Address)
City: ______ State: ______ Zip: ______
County: ______ Phone Number: ______
Operator (if not owner): ______
Qualified Individual(s): (attach additional sheets if more than one)
Name: ______
Position: ______
Work Address: ______
Home Address: ______
Emergency Phone Number: ______
Date of Oil Storage Start-up: ______
Current Operations: ______
Date(s) and Type(s) of Substantial Expansion(s): ______
(Attach additional sheets if necessary)

Emergency Response Information

- use emergency notification phone list[27]
- use facility response equipment list[28]
- use response equipment testing/deployment form[29]

---

[26] 59 FR 34124
[27] 59 FR 34125
[28] 59 FR 34126
[29] 59 FR 34126

- use personnel forms[30]
- prepare facility evacuation plan, including:
  - — facility-wide evacuation
  - — high-risk area evacuation
  - — evacuation route diagram
- qualified individual duties
  - — list of duties of the individuals having full authority to remediate the spill

  [For details refer to 59 FR 34128.]

Hazard Evaluation

- study facility and predict where releases could occur; include:
  - — tanks; form at 59 FR 34129
  - — impoundments, form at 59 FR 34129
- vulnerability analysis, i.e.
  - — water intakes
  - — schools
  - — medical facilities
  - — residential areas
  - — businesses
  - — wetland and sensitive environments
  - — fish and wildlife
  - — lakes and streams
  - — endangered flora and fauna
  - — recreational areas
  - — transportation routes [air, water and land]
  - — utilities
  - — other areas of economic importance
- analysis of potential for an oil spill; consider:
  - — spill history
  - — horizontal range of potential spill
  - — vulnerability to natural disaster
  - — age of tank

---

[30] 59 FR 34126

Facility Reportable Oil Spill History

A reportable oil spill represents a release causing a sheen on Waters of the U.S.[31] "Reportable" in this history means reported and unreported spills. The history must include:

- date of discharge
- list of discharge causes
- material discharged
- amount discharged in gallons
- amount of discharge reaching Waters of the U.S.
- effectiveness and capacity of secondary containment
- clean-up action taken
- storage capacity tank or impoundment from which oil leaked
- action taken to prevent spill reoccurrence
- enforcement action
- effectiveness of monitoring equipment
- methods used to detect each oil spill occurrence

Discharge Scenarios

Describe small, medium, and worst case spill scenarios:

- small case defined as ≤ 2,100 gal [50 bbl] but less than worst case [$S_c \leq 50$ bbl].[32]
- medium case defined as >2,100 gal [50 bbl] but <36,000 gal [858 bbl] or 10% of the largest tank, whichever is less, but less than the worst case [50 bbls < S_m < 858 bbls or 10% of largest tank, whichever is smaller].
- worst case defined as the largest tank within the dike or the sum of the tanks connected together without valves, whichever is greater.

The operator must provide evidence [statement] demonstrating multiple tank installations are not operated as a single tank within the secondary containment area.

The rule provides for development of three [small, medium, and worst case] scenarios. Further, the rule allows for development of one

---

[31] 40 CFR Part 110

[32] 59 FR 34075

[small case] scenario, if the worst case is within the small case range, i.e., ≤50 bbl. This situation appears non-existent within E&P operations, because it would require a 1,000 bbl or 23,810 bbl facility to consist of 50 bbl tanks.

Also, the rule provides for development of two [medium and worst case] scenarios if the worst case is within the medium case, i.e., 50 bbl< S_m < 858 bbl. This situation is possible because two 500 bbl tanks could be present at a 1,000 bbl facility.

For a complex, the operator must calculate the medium spill case using onshore, offshore, non-transportation-, or transportation-related facility as appropriate for the facility and select the largest medium spill.

Discharge Detection System

The operator must discuss and describe the procedures used to detect oil spills during operating hours. The reliability and testing of any automatic equipment must be discussed also. Additionally, the operator must describe the initial response action in the event of a spill. Reference the initial action to the emergency response information.

Plan Implementation

Describe the details of implementing clean-up action planned for small, medium and worst case spills, including:

- emergency plans for spill response
- disposal of collected oil plans

Self-Inspection, Drills/Exercises, and Response Training

Develop programs for response, training, drills/exercises, and facility inspection.

Develop a written procedure for inspecting tanks, dikes, and response equipment including a check list of items to inspect and a form for recording the results of the inspection. Retain inspection records for five years.[33]

---

[33] 59 FR 34132

For response equipment include in inspection:

- inventory
- equipment location
- accessibility

National Preparedness for Response Exercise Program

On July 1, 1994, EPA[34] gave notice of availability of draft guidelines entitled National Preparedness for Response Exercise Program [PREP] from Coast Guard Headquarters. Additionally, EPA indicated a preference for PREP over *ad hoc* exercise programs.

My reading of PREP shows it is well suited to large corporations having a single team serving multiple facilities, because team training at any facility satisfies the requirements at all facilities. On the other hand, small organizations having a few facilities controlled by OPA-90 regulations will find PREP burdensome, costly, and inefficient. I advise small operators subject to OPA-90 to develop a minimum exercise program that is included in the response plan. EPA, RSPA, MMS, and USCG appear to recognize the problem, because on April 20, 1995, they announced a workshop[35] for the purpose of incorporating "industry mentorship" where a small company would participate with a large company as partners in exercises.

DOT RESPONSE PLAN REQUIREMENTS

Jurisdictional Responsibility

On January 5, 1993, RSPA promulgated at 58 FR 244 Response Plan Rule for Onshore Pipelines, 49 CFR Part 194, requiring preparation and submission of a response plan by February 18, 1993.[36]

RSPA defined *onshore facilities* as meaning: "new and existing pipe, right-of-way, and any equipment, facility, or building used in the transportation of oil located in, on, or under, any land within the United States other than submerged land.[37] On July 1, 1994, DOI [MMS

[34] 59 FR 34116
[35] 60 FR 19805
[36] 58 FR 247
[37] 58 FR 254

and USCG], EPA and DOT [RSPA] at 59 FR 34102 published a MOU defining the jurisdictional responsibilities for offshore facilities." The memorandum delegates to DOT [RSPA] responsibility for transportation-related facilities, including pipelines, located landward of the coast line [Figures 3–1 and 3–2]. Refer to EPA response plan requirements for a detailed discussion of the definition of coast line.

For purposes of OPA-90, RSPA has jurisdiction over all DOT pipelines onshore and offshore [shoreward of the SLA[38] coast line deepwater ports and their associated seaward pipelines.][39] No RSPA offshore rule has been promulgated.

I recommend using the onshore rule[40] as guidance in preparing an offshore pipeline response plan.

Response Plan Preparation Criteria

A response plan [Figures 3–1 and 3–3] is required for all RSPA affected pipelines[41] except:

- pipelines 6-5/8 in. or smaller OD *and* 10 mi or less in length *and*
- has not experienced a 1,000 bbl or more release within the previous five years
- has not experienced two or more reportable releases, i.e.,
  - — loss of 50 or more bbl of hazardous liquid or carbon dioxide
  - — escape to the atmosphere of more than five bbl a day of highly volatile liquids[42] within the past five years and
- has electric resistance welded [ERW] pipe manufactured before 1970 and operating at or below 50% of specified minimum yield strength [SMYS]
- is not in proximity to navigable waters, public drinking water intake or environmentally sensitive areas[43] or
- a line section where the operator determines worst case discharge is unlikely to adversely affect within 12 hours of the discharge any navigable waters, public drinking water intake, or environmentally sensitive area.[44]

[38] 49 CFR Part 195
[39] 59 FR 34102
[40] 58 FR 244 dated 5 Jan. 1993
[41] 49 CFR Part 195
[42] 49 CFR §195.50
[43] 49 CFR §194.101 (b)
[44] 49 CFR §194.101

Figure 3–3 DOT response plan flow chart

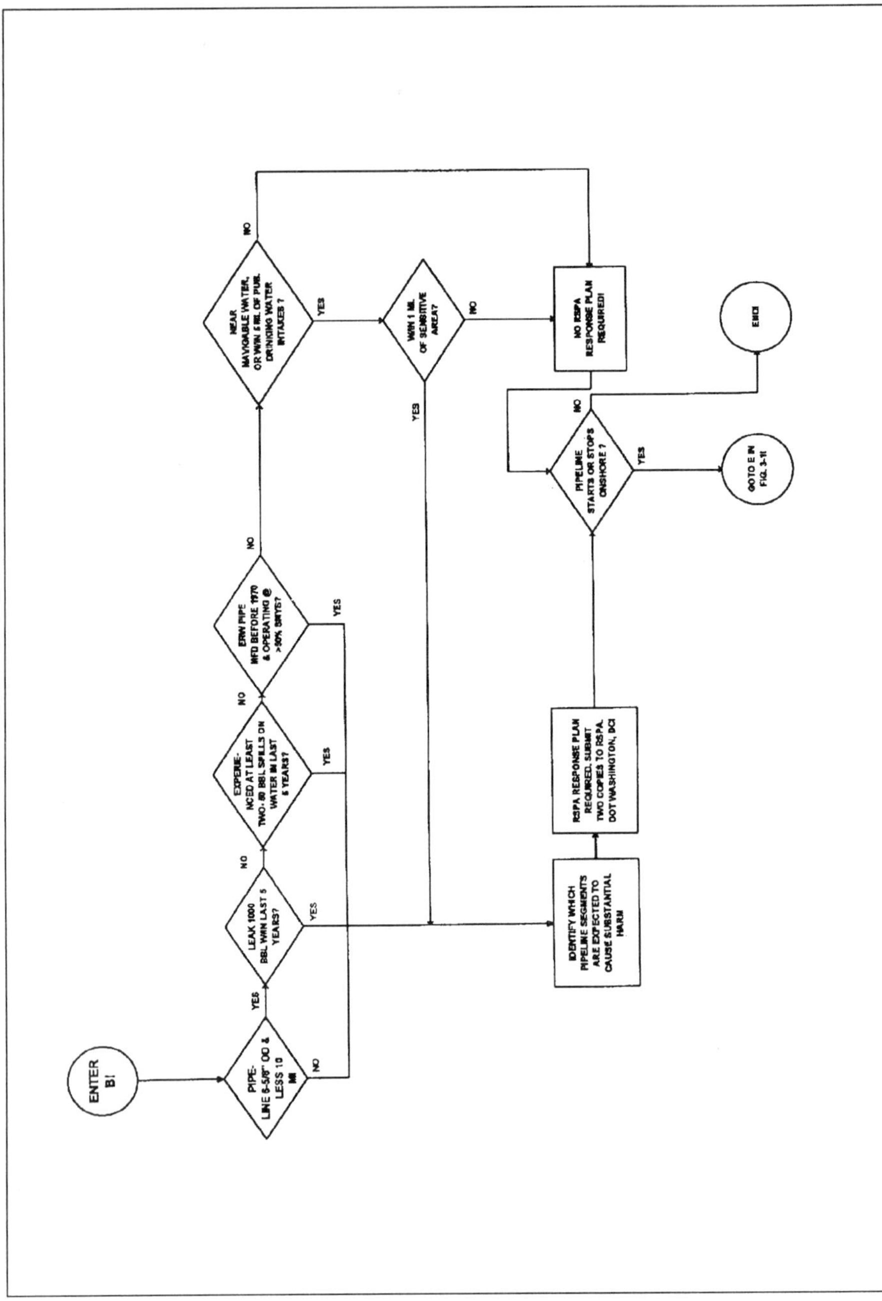

Accordingly, all non-exempted pipelines or line sections located within:

- five mi of potentially affected public water intake *or*
- one mile of a potentially affected environmentally sensitive area

must prepare response plan submission, submit two copies of the response plan to Pipeline Response Plan Officer, Research and Special Programs Administration, 400 7th St. SW, Washington, D.C. 20590-0001.

Figures 3–1 and 3–3 provide a logical method for determining if a facility requires a DOT response plan. Enter at "start" [Figure 3–1] and answer each question "yes" or "no." If the facility represents a DOT facility, Figure 3–3 is exited at "GOTO B." Figure 3–2 is entered at "Enter B" and answer the questions until a determination is made.

Response Plan Requirements

RSPA requires the response plan be written in English and any other language needed to make the plan functional.[45] Each plan must be consistent with the national contingency plan [NCP][46] and each applicable area contingency plan [ACP]. The operator must certify the response plan is consistent with the NCP and applicable to ACP.

Response Plan Content

- name and address of the pipeline operator
- listing and description [by county and state] of each geographic area containing a DOT pipeline for which the operator must plan spill response capability [response zone] i.e., significant and substantial harm.[47]
- an appendix for each response zone must contain:
  - name and address of the pipeline operator
  - description [by county and state] of the geographic area for which the appendix is developed.

[45] 49 CFR §194.107

[46] 40 CFR §300 as revised 15 Sept. 1994 at 59 FR 47384

[47] 49 CFR §194.103

- — notification procedure
- — spill detection and mitigation procedures
- — name, address, and telephone number of the spill response organization [Refer to 49 CFR §194.115 for details.]
- — response activities and response resources
- — names and telephone numbers of federal, state and local agencies having pollution responsibility or support
- — training procedures [Refer to PREP discussion contained in the EPA OPA-90 section.]
- — equipment testing [Refer to PREP discussion contained in the EPA OPA-90 section.]
- — drill types, schedules, and procedures [Refer to PREP discussion contained in the EPA OPA-90 section.]
- — name(s) and telephone number(s) of qualified operator representatives available on a 24-hour basis
- — list of sections for each pipeline within the response zone identified by milepost or survey station number or opera tor designation
- — type of oil and volume of worst case discharge[48]

Worst Case Discharge

The worst case discharge represents the largest volume [bbl] resulting from:

- a pipeline maximum discharge release time [HRS] plus maximum shutdown response time [HRS, based on operator experience or estimate] multiplied by the pipeline flow rate [bbl/hr, based on the maximum daily line capacity] plus the largest line drainage volume for the line section(s) affected [bbl] *or*
- the largest foreseeable release [bbl] within the response zone based on the maximum historical discharge, adjusted for subsequent corrective or preventative action taken *or*
- if the response zone contains one or more breakout tanks, the worst case discharge is the capacity of the largest tank or battery of tanks within a single dike, adjusted for the capacity or size of the secondary containment [The method of adjustment is not specified as of May 1, 1995.][49]

---

[48] 49 CFR §107 and 113

[49] 49 CFR §194.105

State Response Plan

The RSPA rule allows [in lieu of a plan following 49 CFR Part 194] submitting a state response plan if the state rule requires a plan providing equivalent or greater spill protection. The substitute plan must include:

- information summary required by 49 CFR §194.113
- name of qualified individual
- contract for private personnel and equipment capable of responding to the worst case discharge.[50]

USCG RESPONSE PLAN

Requirements and Jurisdictional Responsibilities

On February 5, 1993, at 58 FR 7330, the USCG issued interim final rule[51] for marine transportation-related facilities. The rule applies to:

- fixed onshore facilities capable of transferring to or from a vessel having a capacity of 250 bbl or more and deepwater ports
- mobile onshore facilities [tank trucks and railroad tank cars] used or intended to transfer oil to or from a vessel having a capacity of 250 bbl or more
- marine transportation-related onshore facilities specifically designated by the captain of the port [COTP].[52]

The above facilities are judged by USCG to cause substantial harm to the environment by discharging oil into or onto navigable waters. Marine transportation-related onshore facilities capable of transferring less than 250 bbl of oil are exempt from this rule unless specifically included by the COTP.

USCG defines *onshore facility* at 33 CFR §153.103(o) as meaning "any facility (including but not limited to, motor vehicles and rolling stock) of any kind located in, on, or under any land within the United

---

[50] 49 CFR §§194.109 and 115

[51] 33 CFR Parts 150 and 154

[52] 33 CFR §154.1015

Figure 3–4 Coast Guard response plan flow chart

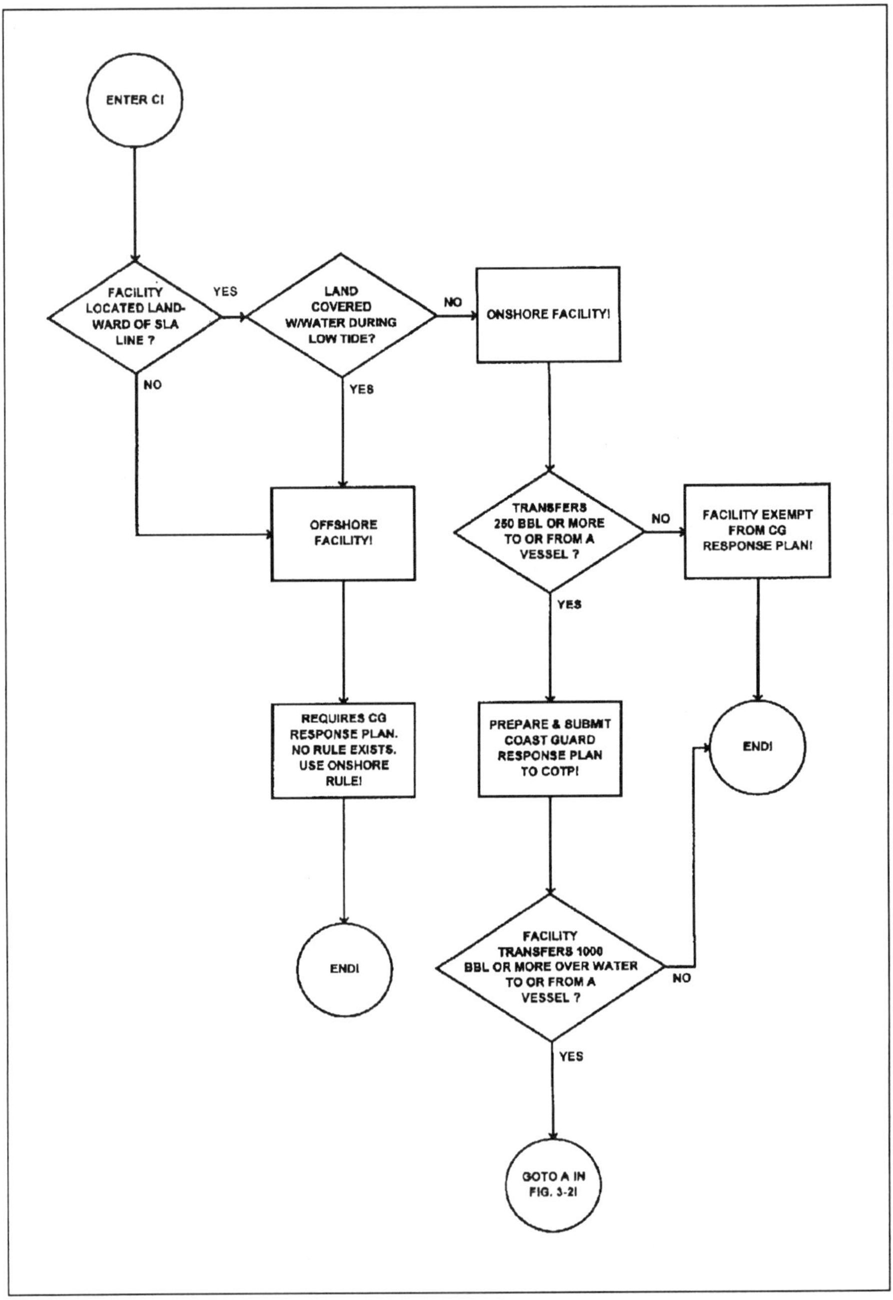

States other than submerged land." Accordingly, 33 CFR Part 154 applies only to dry land facilities. The MOU published on July 1, 1994, at 59 FR 34102 reaffirms this jurisdiction by stating, "DOI retains jurisdiction over facilities...located seaward of the coast [SLA] line...." Thus, the requirement to have an operations manual is not sufficient to require a Coast Guard response plan. Figures 3–1 and 3–4 provide a logical method for determining if a facility requires a Coast Guard response plan. Enter Figure 3–1 at "start" and answer the questions "yes" or "no." If the facility represents a Coast Guard facility, Figure 3–1 is exited at "GOTO C." Enter Figure 3–4 at "Enter C." Answer the questions until a determination is made.

Response Plan Submission

For existing facilities, two copies of the response plan must be submitted to the COPT on or before February 18, 1993. New facilities must submit the response plan to the COTP no less than 60 days prior to initial start-up of the facility. Refer to 33 CFR Part 3 for the applicable COTP.

Response Plan Requirements

Fixed marine transportation-related [MTR] facilities requiring a response plan must comply with 33 CFR §§154.1030–154.1049. These requirements mandate the response plan[53] be:

- written in English
- divided into sections with tabs:

Introduction and Plan Content

- facility name
- facility address, city, county, state, ZIP code, telephone number, and fax number, if so equipped

---

[53] 33 CFR §154.1030

- name, address, and procedures for contacting the owner or operator on a 24-hour basis
- table of contents
- record of changes to record revisions to the plan

Emergency Response Plan Action Plan:

- notification procedures
  A. prioritize list identifying the persons [name, telephone number, and role] to be notified of discharge or threat of a discharge of oil
  B. facility response personnel, spill management team, oil spill removal organization, qualified individuals, and alternates
  C. federal, state and local agencies by name and telephone number. [See Form 3–5.]
  D. spill mitigation procedures
    1. Volumes of persistent and non-persistent oil [54] in the anticipated discharge from the MTR facility
    2 maximum most probable discharge
    3. worst case discharge
    4. prioritized procedures for facility personnel to mitigate or prevent discharge or substantial threat of a discharge. A copy of these procedures must be maintained at a facility operations center. The following scenarios must be discussed:
      a. failure of manifold, loading arm, hose, or other transfer equipment
      b. tank overflow
      c. tank failure
      d. piping failure
      e. piping leak
      f. explosion or fire
      g. equipment failure [pump, valves, relief devices]

---

[54] Non-persistent oil means oil (1) at least 50 percent by volume distills at 340°C [645° F] *and* (2) at least 95 percent by volume distills at 370° C (700° F). All other hydrocarbon oils are persistent oils.

Form 3–4 Information on discharge

Involved Parties

(A) Reporting Party	(B) Suspected Responsible Party
Name	Name
Phone ( )	Phones ( )
Company	Company
Position	Organization Type
Address	Private citizen
Address	Private enterprise
	Public utility
	Local government
	State government
	Federal government
City	City
State	State
Zip	Zip

Were Materials Discharged? Yes____ No____
Calling for Responsible Party? Yes____ No____

Incident Description

Source and/or Cause of Incident

Date Time
Cause

Incident Address/Location Nearest City

Distance from City
Storage Tank Container Type - Above ground (Y__ N__) Below ground (Y__N __)
Unknown
Facility Capacity

Tank Capacity
Latitude Degrees
Longitude Degrees
Mile Post or River Mile

Materials

Discharged Quantity	Unit of Measure	Discharged Material	Quantity in Water

Response Action

Actions Taken to Correct or Mitigate Incident

Impact

Number of Injuries — Number of Fatalities
Were there Evacuations? (Y____N____U____) — Number evacuated
Was there any Damage? (Y____ N____U____) — Damage in Dollars

Additional Information

Any information about the Incident not recorded elsewhere in the report

Caller Notifications

USCG	EPA	State	Other

* IT IS NOT NECESSARY TO WAIT FOR ALL INFORMATION BEFORE CALLING NRC.
NATIONAL RESPONSE CENTER -1-800-424-8802

- listing of equipment and personnel responsibilities to mitigate most probable spill
- response activities
  - describe facility personnel response to a discharge pending arrival of qualified individual[55]
- describe facility and/or corporate organizational structure used to manage response action; include:
  - command and control
  - public information
  - safety
  - liaison with agencies
  - spill operations
  - planning
  - logistical support
  - finance
- oil spill removal organization and team available by contract. The organization and team must have:
  - capability of responding to average most probable discharge, maximum most probable discharge and worst case discharge
  - capability of providing equipment and supplies meeting 33 CFR §§154.1045, 154.1047, or 154.1049; trained person nel for oil spill removal and management for initial seven days of response
- sensitive areas
  - identify areas of environmental sensitivity and economical importance, i.e., wetland, national and state parks, critical habitats for endangered species, wilderness and natural areas, preserves, wildlife areas, scenic and wild rivers, seashore and lakeside recreational areas, and critical biological resources areas.[56]
  - describe response actions anticipated to protect sensitive areas

---

[55] Qualified individual: speaks English; available 24 hrs/day; familiar with implementing response plan; trained in responsibilities of the plan; provided with a document from operator designating the person as a qualified individual; identified in the plan as a qualified individual; delegated authority to activate and engage contractor; act as liaison with federal on-scene coordinator [OSC]; and obligate funds required for response action. The qualified individual may be an organization. Refer to 33 CFR §154.1026 for details

[56] 58 FR 7376

- — provide map of environmentally sensitive areas
- — identify equipment and personnel appropriate[57] to worst case discharge
- — identify equipment and personnel to protect sensitive areas within the distance oils likely will travel and specified times and distances[58]

- disposal plan – describe procedure for disposal of oil and debris.
- training and drills
  A. training procedures
    1. identify training given to each identified individual
    2. identify training given to satisfy hazardous waste operations and emergency response [HAZWOPER] requirements[59]
    3. document training
    4. assure contractor provides and documents HAZWOPER training
  B. drills
    1. use PREP program for drills and exercises
    2. retain drill records for three years
- plan updates
  — review and update response plan annually within one month of the anniversary date of COTP approval
  — send response plan revisions to COTP
  — if plan not revised, submit a letter to COTP so stating

Worst Case Discharge

The worst case discharge for MTR facilities must include:

- loss of the entire contents of the pipeline (leading to the vessel), plus interconnected tanks, plus maximum pumping capacity for the time to:
  — discover the release from the pipe
  — stop pumping operations after discovery

---

[57] Refer to detailed instructions contained.

[58] 58 FR 7358

[59] 29 CFR §1910.120

The above calculated volume is modified by loss factors[60] to account for emulsification and dissipation. Refer to Appendix C[61] for detailed calculation method.

The final solution probably lies in developing cooperatives sponsored by state and/or federal government agencies.

MMS RESPONSE PLAN

Jurisdictional Responsibility

The MOU signed February 3, 1994,[62] provides MMS retain "jurisdiction over facilities, including pipelines, located seaward of the coast [SLA] line, except for deepwater ports and associated seaward pipelines...."

For purposes of a response plan, MMS defines *offshore* to mean: "The area seaward of the line of low water along that portion of the coast which is in direct contact with the open sea and the area seaward of the line marking the limit of inland waters [Barrier Islands]."

Accordingly, OPA-90 as interpreted by MMS, requires a response plan for all non-transportation related facilities capable of releasing oil or hazardous substances to OCS and state waters excluding *state waters inside the barrier islands.*

Response Plan Preparation Criteria

Originally, the MMS contingency plan[63] applied only to the OCS. However, MMS published Interim Final Rule on February 8, 1993,[64] modifying this provision to extend the contingency plan provisions to state waters seaward of the coastline. Also the regulations allow the contingency plan prepared for OCS lease blocks to be used for "facilities in State Waters of the same geographic area" operated by the same operator. Alternatively, the operator can provide a copy of the state

---

[60] 58 FR 7372 and 7373

[61] 58 FR 7366

[62] 59 FR 34102

[63] 30 CFR §250.42

[64] 58 FR 7489

response plan previously submitted to a state for approval, if the transmission includes the following:

- list of offshore facilities and leases included in the plan
- name and address of the state agency receiving the plan
- date plan submitted
- if approved, name of approving agency, date of approval, and copy of approval letter, if issued
- identify rule under which state plan prepared

Further, MMS allows submission of an oil-spill contingency plan if the plan[65] contains:

- summary of available oil-spill trajectory analyses specific to the area of operation
- listing of response equipment and response times along with:
  - A. materials
  - B. support vessels
  - C. procedures to be employed during response to:
    1. worst case discharge
    2. short duration, limited maximum volume releases
- dispersant use plan including:
  - — dispersant inventory
  - — toxicity data
  - — description of type of oil each dispersant is intended to combat
  - — description of dispersant application equipment and procedures
  - — outline of dispersant application approval plan
- provisions for response drills, land inspection, testing, and maintenance of response equipment, [Refer to PREP discussion.]
- early release detection and notification plan including list of:
  - — names, telephone numbers, alternates to responsible people, and names, telephone numbers, addresses of regulatory agencies to be notified in event of a release
- inventory of equipment, materials, and supplies available locally and regionally
- well-defined actions in the event of a release including:
  - — names of response team, designation by name or position of response coordinator, location of response center, and type of communication system

[65] 30 CFR §254.3

- oil disposal plan
- spill monitoring and movement prediction plan
- for Alaska only: ignition plan

The above MMS rule is an Interim Final Rule valid only until February 18, 1995.[66] MMS must either develop a final rule or extend the expiration date. The rule was extended indefinitely.

Financial Responsibility

The most disturbing provision of OPA-90 to E&P operators deals with financial responsibility of offshore operations.[67] The act defines *offshore* for financial responsibility as a facility "in, on or under Waters of the U.S."[68] Accordingly, pipelines crossing lakes or streams, wells drilled under rivers or lakes or wetlands, producing operations in state and OCS waters become "offshore operations" and are subject to $150 million financial responsibility evidence.[69]

The broad nature of the offshore definition causes facilities previously considered onshore to be "offshore." Traditionally, independent oil and gas operators dominate onshore operations. This means small operators must develop a means of providing evidence of $150 million financial responsibility. MMS hearings evidence a reluctance on the part of insurance companies, banks, and other financial institutions to provide fiscal mechanisms, because the $150 million is the lower bound funding and no upper bound is provided by the act.[70] To make matters worse, DOI Solicitor John Leshy, on November 29, 1994, said "OPA unambiguously requires evidence of $150 million in financial responsibility."[71]

Since $150 million of financial responsibility is beyond the resources available to independent operators, two solutions remain, i.e., plug and abandon all independent operator "offshore" sites or petition Congress to amend OPA-90.

---

[66] 58 FR 7489, 8 Feb. 1993

[67] 33 USC 2716

[68] 33 USC 1001 (22)

[69] 33 USC 2716 (c)

[70] 59 FR 34211, 1 July 1994

[71] Leshy, John D. Letter to acting director, Mineral Management Service, 29 November 1994. *Implementation of the Oil Pollution Act of 1990 by Mineral Management Service.*

4 Safe Drinking Water Act

SUMMARY

The *Safe Drinking Water Act* regulates:

- purity of drinking water
- underground injection wells

Providers of water from onsite water wells [regardless of the number of individuals partaking of the water] *should* follow the minimum water purity requirements.[1] A well is defined as an earthen excavation where depth is greater than any surface dimension.

Injection wells are divided into five classes. Class II injection well regulations[2] affect oil and gas produced water disposal and enhanced recovery injection wells. Operators of these wells must have permits either from the EPA or state. Three types of permits are available:

- permit by rule
- individual permits
- area permits

---

1 40 CFR Part 141

2 40 CFR Part 146

Injection is prohibited in any Class II injection well not having a valid permit. Typically, injection well permits expire every five years. Injection pressure, rate, cumulative volume, and mechanical integrity data must be submitted to the State or EPA periodically.

INTRODUCTION

The Safe Drinking Water Act[3] [SDWA], as currently amended, was passed in 1977 as Public Law 95-190. It directly impacts industry in two areas:

- drinking water quality
- injection well control

Wholesome drinking water is assured by periodic sampling and analyses. The SDWA applies to operators providing drinking water to employees and visitors above the threshold of 24 individuals for at least 60 days per year. Regardless of this threshold, I strongly recommend adhering to the minimum coliform, organics, and heavy metals test criteria.

The SDWA imposes strict controls on all injection wells. Typically, E&P operators own or use Class II injection wells for disposal of produced water and injection of oil or gas enhancement fluids. Poor field practices, i.e., lack of injection pressure observations or lack of comparing observations to permit provisions provide grounds for permit cancellation.

Drinking Water Quality

Most E&P facilities do not provide drinking water to employees or the public from onsite water wells. Accordingly, compliance with drinking water requirements of the SDWA[4] are not mandated. However, operators desiring to provide wholesome water for employees and visitors from company owned water wells should voluntarily adhere to the National Primary Drinking Water Standard, even for small work forces.[5]

---

[3] 42 USC 300(f) et. seq.

[4] 40 CFR Parts 141-143

[5] 40 CFR Part 141

Drinking Water Requirements

Definitions

The drinking water requirements of the SDWA apply to public water systems [PWS]. A PWS is one providing water piped at least 60 days per year to 15 service connections, i.e., buildings, shops, laboratories each having a separate line from the water main, or one regularly serving 25 individuals daily at least 60 days per year,[6] Figure 4–1.

Not all PWS are subject to the SDWA provisions. Any PWS that simply distributes and stores water, does not collect and treat water, does not obtain water from another PWS, does not sell water, nor is not classed as a carrier of passengers in interstate commerce is not subject to SDWA regulations.

PWS subject to SDWA regulations are divided into three classifications, Figure 4–1:

- community water systems—providing water to at least 15 service connections used by year-round residents or regularly serving at least 25 year-round residents[7]
- non-community water systems—PWS not providing water to at least 15 service connections or regularly serving an average of at least 25 individuals year around[8]
- non-transient non-community water system—a PWS providing water to at least 25 of the same persons during six months per year[9]

Except at large gas plants or drill sites providing water from onsite wells, few E&P facilities are included within any of the above definitions.

Drinking Water Standards

Nevertheless, minimum requirements governing community water systems need be followed to protect the health of employees and visitors. The regulations include the following standards:

[6] 40 CFR §141.2, Figure 4–1
[7] 40 CFR §141.2
[8] 40 CFR §141.2
[9] 40 CFR §141.2

Figure 4–1 Water systems

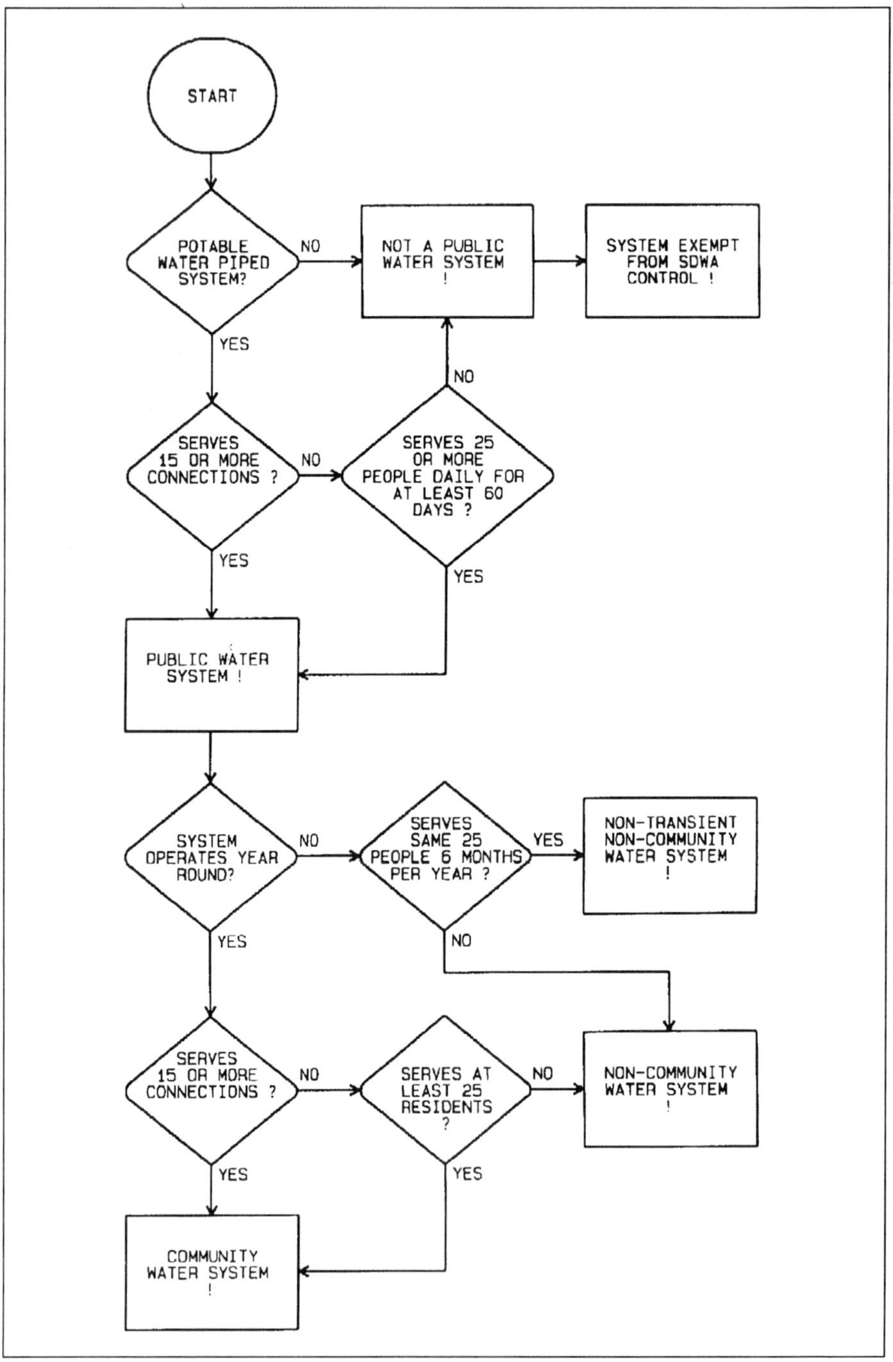

Primary Drinking Water Quality Standards

Primary drinking water quality standards control coliform, metals, and organics, Table 4–1.

Table 4–1 Primary drinking water quality standards

Coliform — No sample should be coliform positive[10]

Heavy Metals[11]

Metals	Maximum Contaminant Level, mg/l
arsenic	0.05
barium	1.0
cadmium	0.010
chromium	0.05
lead	0.05
mercury	0.002
nitrate + nitrite	10.0
selenium	0.01
silver	0.05

Organics[12]

Organics	Maximum Contaminant Level, mg/l
chlorinated hydrocarbons	0.0002-0.1
chlorophenoxys	0.01-0.1
total trihalomethanes	0.10

Secondary Drinking Water Standards

The secondary standards represent non-mandatory guideline concentrations which improve the appearance and taste of drinking wa-

---

[10] 40 CFR §141.63 (a) (2)

[11] 40 CFR §141.11 (b)

[12] 40 CFR §141.12

ter.[13] Accordingly, field water wells need not be analyzed for secondary components.

Drinking Water Sampling Frequency

The EPA requires sampling on a prescribed schedule to assure wholesome drinking water:

- coliform[14] — 25-1000 persons served, one sample/month
- heavy metals[15] — once/year
- organics[16] — once/year

Analytical Methods

The EPA specifies analytical methods[17] for determining the purity of drinking water. A laboratory must be required to utilize these methods when analyzing drinking water. The method must be recorded on the analytical report.

Notification Requirements

SDWA Section 1414(c) requires the owner or operator of a PWS give notice to persons served by the system of the failure of the water to comply with any maximum contaminant level [MCL] or monitoring requirement.[18] For exceedance of MCL, notice consists of :

- publication in a daily newspaper
- publication in a weekly newspaper, if the area is not served by a daily paper, but at least within 14 days of non-compliance
- direct mail or hand delivery within 45 days

---

[13] 40 CFR §143.1
[14] 40 CFR §141.21
[15] 40 CFR §142.23
[16] 40 CFR §142.24
[17] 40 CFR §§141.89 and 136.3
[18] 40 CFR §141.32, 52 FR 41534

The notice must contain:

- explanation of the violation, adverse health effects, and corrective action taken
- telephone number of water provider

Refer to 40 CFR §141.32 for detailed information regarding noncompliance with organic MCL's.

UNDERGROUND INJECTION CONTROL

The EPA promulgated underground injection control [UIC] technical regulations as 40 CFR Part 146 in 1980.[19] Subsequently, they promulgated 40 CFR Part 144 in 1983[20] dealing with UIC program requirements and 40 CFR Part 147 in 1984[21] dealing with EPA-administered state programs.

SDWA Section 1425 provides for utilization of existing state UIC programs where such programs existed prior to adoption of the SDWA. This allowed states the option of continuing existing state permits as permits by rule. Accordingly, most oil and gas producing state programs were not interrupted by the promulgation of 40 CFR Part 146.

Underground Injection Control Definitions

The act and implementing rules provided definitions and terms not necessarily familiar to E&P operators:

- area of review — the area surrounding an *injection well*
- contaminant — any physical, chemical, biological, or radioactive substance in water
- director — the regional administrator [EPA], state director or tribal director responsible for the UIC activities within a state
- exempted acquifer — a formation containing an underground source of drinking water [USDW], but does not or will not serve as a USDW[22]

---

[19] 45 FR 42500
[20] 48 FR 14189
[21] 49 FR 20197
[22] 40 CFR §146.4

- sole source aquifer — an aquifer designated by the EPA administrator as the sole source of ground water for an area
- underground source of drinking water — an aquifer that (1) supplies any PWS; (2) contains sufficient quantity of ground water to supply a public water system, and (a) supplies drinking water for human consumption *or* (b) contains < 10,000 mg/l total dissolved solids [TDS] *and* (c) is not an exempted acquifer
- well — a bored, drilled, or driven shaft or dug hole where the depth is greater than any surface dimension
- well monitoring — the measurement by instruments or laboratory methods of the quality of well water

Injection Well Classifications

The EPA defines five classes of injection wells:[23]

- **Class I** — Wells injecting hazardous or non-hazardous waste[24] *below* USDW, [<10,000 mg/l TDS, 40 CFR § 146.3] within 1/4 mi of a water well which provides USDW[25]
- **Class II** — Wells injecting fluids brought to the surface with conventional oil and gas production, fluids used for enhanced oil or gas recovery, fluids from in-field gas plant [i.e., produced water, cooling tower blowdown and/or boiler blowdown], and wells injecting hydrocarbons which are liquid at room temperature and pressure, i.e. strategic reserve injection wells[26]
- **Class III** — Wells injecting solvents [liquids or gases] for the purpose of solution mining [uranium, sulfur, metals, salt, or potash][27]
- **Class IV** — Wells injecting hazardous or radioactive waste into or above formations containing USDW within 1/4 mile of the well.[28] [These wells are prohibited.[29]]

---

23 40 CFR §146.5
24 40 CFR §261.21-33
25 40 CFR §144.6 (e)
26 40 CFR §144.6 (b)
27 40 CFR §144.6 (c)
28 40 CFR §144.6 (d)
29 40 CFR §144.13

- **Class V** — Wells not falling into any of the above classes, i.e., septic tank leaching field drainage wells, liquefied petroleum gas [LPG] storage wells, air conditioning return flow wells[30]

Class II Injection Well Permits

Three types of UIC permits are available to Class II well operators:

- permit by rule
- individual permit
- area permit

Permit by Rule

Any state with a UIC permit program in operation prior to passage of the SDWA and granted UIC primacy [authority to operate the UIC permit program] could continue existing state permits for disposal wells in lieu of issuing a an individual or area permit for the UIC well. The extension of an existing state UIC permit is classed as a permit by rule. These permits by rule for Class II wells, except enhanced recovery injection wells, were designed to be replaced by an individual permit within five years of approval of the state UIC program.[31] As a practical matter, not all of the permits by rule have been replaced as of, May 1, 1995. Accordingly, the permit by rule remains in force if the operator applies for a permit and continues to comply with the permit by rule provisions.[32]

Individual Permit

Class II permits are divided into two groups:

- disposal wells
- enhanced recovery injection wells

[30] 40 CFR §144.6 (e); A number of states traditionally included jug injection/recovery wells under the jurisdiction of the oil and gas commission. Accordingly, these states continue classing LPG jug wells as Class II wells.
[31] 40 CFR §144.21 (a) (3) (ii)
[32] 5 UCS 558 (c) (2)

Permits for both types of wells have a term of five years. However, if a permit application was filed in a timely fashion and the operations comply with the expiring permit, the permit remains in effect. If a permit application is not filed on time or is judged not complete, injection must cease.[33] Interestingly, the EPA provides no rule for UIC permit renewal. I suggest submitting a renewal permit application nine months before the existing permit expires.

An injection well [except an enhanced recovery injection well] authorized by rule must apply for an individual permit if requested by the director. For EPA-administered programs [non-primacy states and certain Indian lands], the regional administrator may require the operator to submit an application only after the owner or operator is notified in writing.[34]

Permit applications for all UIC permits except Class II wells must be signed by a corporate officer or general partner. Class II permit applications must be signed either by a corporate officer, general partner, or a duly authorized representative, i.e., authorized in writing by a corporate officer or general partner. The written authorization must be submitted to the director.[35]

Additionally, any document, permit application, report, or inventory must be certified as follows:

> I certify under penalty of law that this document and all attachments were prepared under my direction or supervision in accordance with a system designed to assure that qualified personnel properly gather and evaluate the information submitted. Based on my inquiry of the person or persons who manage the system, or those persons directly responsible for gathering the information, the information submitted is, to the best of my knowledge and belief, true, accurate, and complete. I am aware that there are significant penalties for submitting false information, including the possibility of fine and imprisonment for knowing violations.

Area Permits

Area UIC permits apply to injection wells within the same field, facility, reservoir, or project operated by a single operator.[36] These

[33] 40 CFR §144.11

[34] 40 CFR §144.25

[35] 40 CFR §144.32

[36] 40 CFR §144.33

wells may not inject hazardous waste. Typically, area permits apply to unitized operations or large water flood operations.

AREA OF REVIEW

The regulations[37] require reviewing the area surrounding an injection well for possible sources of migration from the well or injection zone into USDW. Two means of determining the area of review are allowed:

- zone of endangering influence
- fixed radius

The application of the area of review depends on type of permit desired, i.e., individual permit[38] or area permit.[39]

For individual permits, the area of review consists of the lateral distance of pressure influence caused by the injection. In the case of area permits the area of review consists of the project area — field boundary — plus the lateral distance of pressure influence.

Within the area of review, the operator must demonstrate all wells penetrating the injection zone are not leakage paths by which injected and formation fluids may enter USDW.[40] For existing Class II wells, mechanical integrity of the well can be demonstrated by using of cementing records to show adequate cement to prevent fluid migration.[41] If, in the opinion of the director, insufficient cement exists or well integrity is not demonstrated within the area of review, corrective action is required.[42]

The director may disregard the area of review and corrective action when reviewing injection well permit applications for Class II wells. However, experience indicates the director will require area of review and corrective action submissions.

Zone of Endangering Influence

The zone of endangering influence is determined by a formula, Equation 4-1, based on an infinite size reservoir having only one

[37] 40 CFR §146.6
[38] 40 CFR §122.38
[39] 40 CFR §122.39
[40] 40 CFR §144.55
[41] 40 CFR §146.8
[42] 40 CFR §146.07

injection well penetrating the full depth of the injection zone.[43] This is not a realistic model, because the formula provides extremely conservative results that can impose unrealistic hardships on the well operator. Accordingly, the fixed radius method is recommended for area of review determinations.

$$r = \frac{2.25KHt}{S10^{x}} \ {}^{1}\!/_{2} \qquad \text{Equation 4-1}$$

where:

$$x = \frac{4\pi KH\,(h_w - h_{bo}{*}Sg)}{2.3Q}$$

r = radius of endangering influence from the well [length]
K = hydraulic conductivity of the injection zone formation [length/time]
H = thickness of injection zone [length]
t = time of injection [time]
S = storage coefficient [dimensionless]
Q = injection rate [volume/time]
h_{bo} = hydrostatic head of USDW [length]
measured from the base of the lowermost USDW
h_w = hydrostatic head of USDW [length] measured from the base of the lowermost USDW
Sg = specific gravity of injection zone fluid [dimensionless]
π = 3.14159

The radius of endangering influence [r] is the lateral distance in determining area of review. The radius may be less than 1/4 mi when determined by use of Equation 4-1.

Fixed Radius

The regulations allow use of 1/4 mi fixed radius around the injection well or project area. In the case of an area permit [same well design for a number of injection wells serving a field], the 1/4 mi radius applies to a 1/4 mi wide area circumscribing the area, i.e., 1/4 mi on *any* side of the field boundary.

---

43 40 CFR §146.6 (a) (2)

The regulations[44] also require a demonstration of the mechanical integrity of the well.

Mechanical Integrity Test

The EPA requires Class II wells, except packerless completions, to perform a mechanical integrity test [MIT] every five years. The MIT consists of two parts[45] demonstrating:

- no significant leak in the tubing, packer, and/or casing
- no significant fluid movement into USDW through vertical channels adjacent to the injection well bore [Note the word "fluid" is not limited to injected fluid. Accordingly, fluids moving from a penetrated formation must be considered.]

Packer-Equipped Wells

The EPA provides a number of means of determining mechanical integrity, Table 4–2.[46]

Table 4–2 Injection well mechanical integrity

Well Class	Method
I, II, III	Monitoring positive annulus pressure [58 FR 63890]
I, II, III	Pressure test of annulus and tubing
Existing II	Monitoring records showing absence of significant change in the relationship between injection pressure and injection rate
Existing II without long string	Director's discretion

[44] 40 CFR §146.8
[45] 40 CFR §146.8
[46] 40 CFR §146.8

Packerless Completions

The UIC rules for Class II wells do not require packers. However, most of the states require packers except in tubingless completions, steam injection wells, and storage injection/withdrawal wells. However, 40 CFR §146.8(b)(3)(j) requires operators of these wells to conduct a pressure test when the well is placed in service and when the well is taken out of service for reconditioning.

Significant Fluid Movement Mechanical Integrity Test

Additionally, the EPA specifies monitoring methods for determining the absence of significant fluid movement between USDW, Table 4–3.

Table 4–3 Injection well integrity

Well Class	Method
All	Temperature log
All	Noise log
All	Oxygen activation log[47]
All	Tracer survey
Existing II	Cementing records

The most obvious and practical method for determining integrity of the tubing, tubing-casing annulus, and casing is by means of a pressure test. Typically, states granted primacy under the SDWA establish a test procedure specifying a maximum pressure and an allowable fall-off percentage as a function of time. The procedures vary from state to state.

Another method of mechanical integrity testing, which has the advantage of real time, is monitoring the casing-tubing annulus. This technique requires maintaining a positive annulus pressure and comparing the correlating injection rate with injection pressure. Basically, this method is ambient temperature sensitive. However, volume/psi

[47] 56 FR 4063; American Petroleum Institute challenged this method in court.

of injection pressure changes only slight with temperature on a daily basis. Therefore, if a substantial change occurs in barrels/psi, it represents an indication of a substantial change in operation. If the annulus pressure and injection rate remain nearly constant, the well integrity is satisfactory and the operator can assume the change results from formation changes.

Future Mechanical Integrity Test Rule

The EPA believes conducting a MIT once every five years on all wells does not provide adequate assurance of leak-free operation for wells not cemented from the surface to below the USDW.

Accordingly, a system is being developed for decreased intervals between MITs as a function of the well completion configuration.[48] The EPA is considering rating injection wells on the number of levels of protection offered. Levels would be defined as:

- casing cemented from surface to below USDW
- well casing from surface to the injection zone
- well completed with tubing and packer

Thus, a maximum of three layers of protection are possible. If the well has three layers of protection, MIT would be required every five years. If the well has two layers of protection, MIT would be required every three years. If the well has one layer of protection, MIT would be required every year.

This means new wells completed to current standards [below] would be tested less frequently than most existing wells.

Well Completion Requirements

All new Class II injection wells must be cased and cemented to prevent movement of fluids into or between USDW.[49] These requirements do not apply to new, existing wells, or newly converted Class II wells located in *existing fields if*:

- state regulatory controls for casing and cementing existed for those wells at the time of drilling and those wells complied with the controls

[48] 59 FR 58200; 14 Nov. 1994

[49] 40 CFR §146.22 (b)

- injection will not result in movement of fluids into USDW so as to create a significant health risk to persons [Note: The EPA is interpreting this provision so as to control movement of fluids between zones not containing USDW.[50]

Further, new Class II injection wells must be logged and/or tested for USDW and confining zones. A descriptive interpretive report of the logs/tests must be submitted to the director.[51] For a surface hole, the logs and tests must include at least:

- hole deviations, if the well is drilled using a pilot hole and subsequently reamed to full size
- any other logs or tests as needed by the director; typically electric log and caliper log

For an intermediate and long string hole, the logs and tests can include:

- electric, porosity, and gamma ray logs
- fracture finder log
- cement bond, temperature, or density log after setting and cementing casing

Injection Formation Information

The following information must be determined for new Class II wells:[52]

- formation fluid pressure
- estimated formation fracture pressure
- physical and chemical characteristics of the injection zone

Well Operating Requirements

Class II injection wells must operate at injection pressures calculated not to exceed pressures that would initiate fractures or propa-

[50] 40 CFR §146.22 (c)
[51] 40 CFR §146.22 (f)
[52] 40 CFR §146.22 (g)

gate existing fractures in *confining zones* of USDW.[53] Also, injection pressure must be limited so injection or formation fluids do not enter USDW.[54] Further, injection between casing protecting USDW and the well bore [hole] is prohibited. Additionally, the EPA imposes a maximum injection pressure on fields where states do not have primacy, Table 4–4. Typically, the maximum injection pressure is below fracturing pressure of the injection zones.[55]

Table 4–4 Federally operated UIC programs for Class II wells

State	Programs
Alabama	Indian lands Class II wells
Alaska	Indian lands Class II wells
Arkansas	Indian lands Class II wells
California	Indian lands Class II wells
Colorado	All Class II wells
Florida	All Class II wells
Idaho	Indian lands Class II wells
Illinois	Indian lands Class II wells
Indiana	All Class II wells
Kansas	Indian lands Class II wells
Kentucky	All Class II wells
Louisiana	Indian lands Class II wells
Michigan	All Class II wells
Mississippi	All Class II wells
Montana	All Class II wells
Nebraska	Indian lands Class II wells
Nevada	Indian lands Class II wells
New Mexico	Indian lands Class II wells
New York	All Class II wells
North Dakota	Indian lands Class II wells
Ohio	Indian lands Class II wells
Oklahoma	Indian lands Class II wells except the Five Civilized Tribes
Pennsylvania	All Class II wells
North Dakota	Indian lands Class II wells
Tennessee	All Class II wells
Texas	Indian lands Class II wells
Utah	Indian lands Class II wells
Washington	Indian lands Class II wells
Wyoming	Indian land Class II wells
Osage Mineral Reserve	All class II wells

[53] 40 CFR §146.23 (a)

[54] 40 CFR §146.23 (a)

[55] 40 CFR Part 147

Monitoring Requirements

Injection wells must be monitored by *observing* the:

- injection pressure
- injection rate
- cumulative volume at least:
  - — weekly for produced water disposal wells[56]
  - — monthly for enhanced recovery injection wells [water flood, continuous steam flood, polymer flood, insitu combustion wells][57]
  - — daily for liquid hydrocarbon storage injection well while injecting or withdrawing hydrocarbons[58]
  - — daily for cyclic steam injection wells[59]

Additionally, the operator must record one observation of injection pressure, flow rate, and cumulative volume at intervals *not exceeding 30 days*.[60] All monitoring records must be maintained until the next permit review, i.e., five years. Also, the operator must identify the types of tests and methods used during pressure, rate, and volume monitoring.[61]

Monitoring may be performed at a common manifold, if the injection system is so designed.[62] Finally, the nature of the injected fluid must be monitored with sufficient frequency to yield data representable of fluid characteristics, i.e., every time a new well is introduced into the system.[63]

Reporting Requirements

The operator must report annually the monitoring results to the director. The report must include cumulative volume and average injection pressure and rate for each month during the calendar year.[64]

[56] 40 CFR §146.23 (b) (2) (i)
[57] 40 CFR §146.23 (b) (2) (ii)
[58] 40 CFR §146.23 (b) (2) (iii)
[59] 40 CFR §146.23 (b) (2) (iv)
[60] 40 CFR §146.23 (b) (2) and 40 CFR §146.23 (b) (4)
[61] 40 CFR §144.52 (a) (5)
[62] 40 CFR §146.23 (b) (5)
[63] 40 CFR §144.26 (b) (1)
[64] 40 CFR §144.26 (c) (1)

The report should include a narrative summary of changes in injected fluid characteristics or sources. Previously reported information, i.e., lack of mechanical integrity should be included by reference. If the pressure and rate data are taken from a common manifold, the data may be submitted in that form.

Class V Wells

A few Class V wells, typically septic tank drainage field wells, are located at field offices or doghouses. Currently, no permit is required for class V wells. However, the EPA requires Class V wells be inventoried within three years of approval of a state program. The information must include:[65]

- facility name and location
- name and address of legal contact
- ownership of the well
- nature and type of injection
- operating status of the well

Also, the inventory must include:[66]

- well completion features
- nature and volume of injected fluids
- assessment of contamination potential of the Class V well based on state hydrogeological data
- assessment of available corrective alternatives, where appropriate, and their environmental and economical consequences

EPA–Administered Injection Well Programs

Some states elected not to seek UIC primacy, i.e., right to administer the program at the state level for injection wells under the SDWA. These state programs and Indian land UIC programs are administered by the EPA, Table 4–4.

---

[65] 40 CFR §144.26 (a)

[66] 40 CFR §146.52 (b)

For most programs administered by the EPA, a maximum injection pressure is specified by a highly simplified and conservative formula in the form of:

Pm = [(Pg-Pw)(Sg)]d

where: Pm = injection pressure at wellhead, psig
Pg = assumed formation fracture gradient, psi/ft of depth
Pw = pressure gradient of fresh water, 0.433 psi/ft
Sg = specific gravity of the injected fluid
d = depth to top of injection perforations, ft

This formula ignores frictional losses in the injection string. Accordingly, the maximum injection pressure calculated from the formula is conservatively low. Refer to 40 CFR §147.103 through 147.2153 for applicable formulas for states included in Table 4-4. For primacy wells, the EPA relies on the state-administered program for determining maximum injection pressures.

EPA Program Requirements

The EPA divides the UIC regulations into two basic parts:

- EPA- administered state programs, and
- EPA -administered Indian lands programs

The state programs are detailed in 40 CFR Part 147 by states. The Indian lands are subdivided into three parts:

- Osage Mineral Reserve[67]
- lands of the Navajo, Ute Mountain Ute, and all of the New Mexico tribes[68]
- lands of certain Oklahoma Indian tribes[69]

[67] 40 CFR §147.2901 147.2929

[68] 40 CFR §147.3000-147, Subpart HHH, Appendix A

[69] 40 CFR §§147.3100-147.3109

5 Toxic Substances Control Act

SUMMARY

The *Toxic Substance Control Act* [TSCA] became effective January 1, 1977, and mandates controls of all chemical substances, i.e., any organic or inorganic having a particular molecular identity[1] and their mixtures manufactured, imported, processed, distributed, or used for commercial purposes within the United States. TSCA[2] fills the regulatory gap left by the *Federal Food, Drug, and Cosmetic Act* and the *Federal Insecticide, Fungicide, and Rodenticide Act*. Accordingly, substances regulated under those acts are not regulated under TSCA.

INTRODUCTION

TSCA was initiated as a means of controlling chemical substances that could have an adverse effect on people and the environment.

Initially, the petroleum extraction industry assumed TSCA applied only to chemical manufacturers. However, in preparing guidance documents and implementing regulations, the EPA expanded the definitions and scope of TSCA to include almost any industry. The redefinition of "manufactured" includes "produced." Thus, manufacturing of chemical substances and mixtures includes *production* of

1 40 CFR §720.3 (e)

2 15 USC Section 2601 et seq.

crude oil, condensate, natural gases, liquefied petroleum gases, sulfur, and coal, bringing TSCA controls on E&P operators.

Many of the TSCA controls on manufacturers, however, do not apply to E&P operations. Accordingly, care must be exercised to avoid violating the provisions of TSCA. The following provisions can apply to E&P operations:

- Section 6 — Regulated hazardous chemical substances and mixtures
- Section 8(a) — records for master inventory of chemicals
- Section 8(b) — master inventory
- Section 8(c) — records of allegations
- Section 8(d) — reporting of health and safety data
- Section 8(e) — reporting of substantial risk

REGULATED HAZARDOUS CHEMICAL SUBSTANCES AND MIXTURES

TSCA Section 6 authorizes the EPA to regulate substances manufactured, processed, distributed in commerce, used or disposed of and judged by the EPA to represent an unnecessary risk to human health or the environment. Accordingly, the EPA controls hexavalent chromium use and disposal associated with industrial cooling tower water treatment,[3] polychlorinated biphenyls,[4] and asbestos.[5]

Hexavalant Chromium Treatment

TSCA Section 6 prohibits distribution in commerce of hexavalent chromium-based water treatment chemicals[6] and commercial use of hexavalent chromium-based water treatment chemicals in non-industrial *comfort cooling towers*. These prohibitions became effective February 20, 1990, and May 18, 1990, respectively.

Some E&P operators provide air conditioning systems having open or closed loop water cooling systems. The use of hexavalent chromium-based water treatment chemicals [sodium dichromate] in these

[3] 40 CFR Part 749
[4] 40 CFR Part 761
[5] 40 CFR Part 763
[6] 40 CFR §749.68 (e)

systems is prohibited. However, the prohibition does not include distribution and use of hexavalent chemicals in industrial cooling towers.[7] The distributor of such chemicals for industrial use is required to maintain records of all shipments for industrial purposes.

Polychlorinated Biphenyls

TSCA Section 6[8] severely limits use of polychlorinated biphenyls [PCB] and rigidly controls disposal of all PCB and PCB articles containing 50 mg/kg or more PCB.

PCB Definitions

When added to hydrocarbon oil, PCB acts as a fire retardant.

- *PCB* is a persistent chlorinated biphenyl molecule [material] that is expressly controlled under TSCA Section 6.[9]
- An *article* is a formed item which functions because of that form, i.e., a transformer.

In E&P operations, several articles are used:

- transformers
- capacitors
- heat circulating systems

All of these articles can contain PCB in varying concentrations.

Transformers

Electric transformers are manufactured as dry or liquid-filled types. Dry transformers are not controlled by TSCA. Liquid-filled transformers can be classed as:

- PCB transformers
- PCB contaminated transformers
- Non-PCB transformers

---

7 40 CFR §749.68 (e) (3)

8 40 CFR Part 761

9 40 CFR §761.3

A liquid-filled transformer is defined[10] as a PCB transformer when the PCB concentration in the dielectric fluid is 500 ppm [mg/kg] or higher [PCB >500 ppm]. A liquid-filled transformer is defined as a PCB contaminated transformer when the PCB concentration in the dielectric fluid is less than 500 ppm but not less than 50 ppm [50≤PCB <500 ppm]. A liquid-filled transformer is a non-PCB transformer when the PCB concentration is less than 50 ppm [PCB <50 ppm].

Any transformer not identified by dielectric fluid analysis or name plate data as a PCB contaminated or Non-PCB transformer must be classed as a PCB transformer.[11] If the name plate or purchase order indicates the transformer is oil filled, the transformer can be assumed to be a PCB contaminated transformer.[12] Liquid-filled transformers manufactured after January 1, 1979, and not marked PCB transformer are non-PCB transformers.

Capacitors

Any liquid-filled capacitor manufactured before January 1, 1979, is classed as a PCB capacitor. Any liquid-filled capacitor manufactured on or after January 1, 1979, and not labeled PCB is a non-PCB capacitor.[13]

Capacitors are classed as small and large. A small capacitor satisfies all of the following criteria.[14]

- Contain less than 3 lbs of PCB
- Weighs less than 9 lbs total weight
- Volume less than 200 cu in.

All other capacitors are large capacitors.

Heat Medium Fluids

Heat medium systems were mixed with PCB to make the oil fire resistant. These fluids are classed as follows:

- PCB fluid, PCB ≥500 ppm

[10] 40 CFR §761.3
[11] 40 CFR §761.3
[12] 40 CFR §761.3
[13] 15 USC 2605 (e)
[14] 40 CFR §761.3

- PCB contaminated fluid, 50 ≤PCB <500 ppm
- Non-PCB fluid, PCB <50 ppm

Disposal of these fluids is strictly controlled as discussed below.

PCB Usage

The EPA prohibits use of PCB electrical equipment threatening food for humans and feed for animals after October 1, 1985, for transformers and after October 1, 1988, for large capacitors, unless contained in an electrical substation.[15] A precise definition of an electrical substation is not available. In my opinion, an electrical substation is a limited access fenced area containing transformers and/or capacitors. The substation need not have an impervious surface. However, I expect the fenced area to be large enough to nominally prevent PCB leaks from contacting food and feed.

For large liquid-filled capacitors, substation has been defined by the EPA in a letter to the API as "any fenced area." This interpretation results in the typical fenced lease or fee property being classed a capacitor substation. As a practical matter, this broad definition provides little additional exposure, because few large liquid capacitors are present in E&P operations.

PCB Transformer Usage

PCB transformers may be used in restricted access substations under specific guidelines.[16]

Inspection

Each PCB transformer in use must be inspected on a schedule dictated by the location and risk reduction measures employed.[17]

- Visually inspect each PCB transformer in use or in storage for reuse at any time during every calendar quarter. A 30-day period must insue between inspections.[18]

---

[15] 40 CFR 761.30 (a) (1) & 761.30 (l) (1)

[16] 40 CFR §761.30

[17] 40 CFR §761.30 (a)

[18] 40 CFR §761.30 (a) (1) (ix)

- A reduced visual inspection frequency of once every 12 months applies if the transformer is located on an impervious, undrained secondary containment having a capacity of 100% of the total dielectric fluid within the containment[19] or if PCB concentration is <60,000 ppm[20]
- An increased visual inspection frequency of once per week applies to PCB transformers threatening food or feed.[21]

If a PCB transformer leak is observed, i.e., liquid on case exterior, repair or replace the transformer. Initiate clean-up within 48 hours. Inspect transformer and containment daily to verify containment of PCB liquids.[22]

Records

The EPA requires the following records be maintained for a period of three years for all PCB transformers stored or in use:[23]

- storage or use location
- date of inspections
- date of each leak discovery
- name of persons making inspections
- location of leaks discovered
- estimate of dielectric fluid released
- date of clean-up, containment, and/or replacement
- results of containment and daily inspections required for uncorrected active leaks

Transformer Servicing

Liquid-filled transformers can be serviced and reused under controlled conditions established by 40 CFR § 761.30.

PCB Transformers

Service [refill] with dielectric fluids containing any concentration of PCB.[24] However, repair effected by coil removal is prohibited.[25] Any

[19] 40 CFR §761.30 (a) (1) (xi)
[20] 40 CFR 761.30 (a) (1) (xi)
[21] 40 CFR §761.30 (a) (1) (xiv)
[22] 40 CFR §761.30 (a) (1) (x)
[23] 40 CFR §761.30 (a) (1) (xii)
[24] 40 CFR §761.30 (a) (2) (ii)
[25] 40 CFR §761.30 (a) (2) (ii)

fluid removed from a PCB transformer must be used in PCB transformers or disposed of in accordance with 40 CFR Part 761.60.

Bulk PCB fluids containing < 500 ppm cannot be mixed with fluid containing ≥ 500 ppm PCB.[26]

PCB Contaminated Transformers

Service [refill] with dielectric fluids containing < 500 ppm PCB [preferable service is with mineral oil].[27]

PCB Transformer Registration

All PCB transformers, including those in storage for reuse, must be registered with the local fire department.[28] Required information includes:

- location of PCB transformer in use or in storage for reuse; physical location of PCB transformer at the site
- dielectric fluid constituent, i.e. PCB, mineral oil, silicone oil
- name and telephone number of company contact in event of emergency

Additionally, PCB transformers in use in or near commercial buildings must register the presence of PCB with the building owner.[29] The information must include:

- specific location of transformer
- dielectric constituent
- type and voltage of transformer

Capacitor Usage

Liquid-filled capacitors manufactured before January 1, 1979, are assumed to be PCB capacitors. Use is prohibited after October 1, 1988, if they threaten food and feed.[30] However, PCB capacitors are not prohibited if located at a controlled access substation.[31]

---

[26] 40 CFR §761.30 (a) (2) (iv)
[27] 40 CFR §761.30 (a) (2) (i)
[28] 40 CFR §761.30 (a) (1) (iii) (C) (2) (iv)
[29] 40 CFR §761.30 (a) (1) (vii)
[30] 40 CFR §761.30 (l) (1) (i)
[31] 40 CFR §761.30 (b) (1) (ii)

PCB Marking

Marking must follow the requirements of 40 CFR §§ 761.40 and 761.45.

The EPA requires the marking [labeling] of the following PCB items:

- PCB containers[32]
- PCB transformers[33]
- PCB large capacitors[34]
- equipment containing PCB transformers or large capacitors at the time of the equipment removal[35]
- electrical motors using PCB coolants[36]
- PCB articles containing any of the above items[37]
- storage areas used for PCB or PCB items for disposal[38]
- transport vehicles, all four sides, when loaded with PCB containers holding > 45 kg [99.4 lbs] of PCB in the liquid phase *or* when loaded with one or more PCB transformers[39]
- vault doors, machinery room doors, fences, hallways, and means of ingress or egress to PCB transformers[40]

Storage for Disposal

PCB contaminated electrical equipment which has been drained of free-flowing dielectric fluid is not subject to the storage for disposal requirements.[41] However, all liquid-filled PCB electrical equipment is subject to the rule. Two types of storage are recognized by the regulations:

- temporary storage [30 days or less]
- specification storage [one year or less]

---

32 40 CFR §761.30 (a) (1)
33 40 CFR §761.40 (a) (2)
34 40 CFR §761.40 (a) (3)
35 40 CFR §761.40 (a) (4) (5)
36 40 CFR §761.40 (a) (6)
37 40 CFR §761.40 (a) (9)
38 40 CFR §761.40 (a) (10)
39 57 FR 45446
40 40 CFR §761.40 (j) (1)
41 40 CFR §761.65 (c) (2)

Temporary Storage

- EPA imposes no physical or location storage requirements.[42]
- Label notation attached to the container of a leaking item must indicate date of removal from service[43] and state contents do *not exceed 500 ppm PCB*.[44]

The following non-leaking PCB or specially packed articles, equipment, or containers may be placed in temporary storage:[45]

- containers holding leaking PCB articles packed in sorbents[46]
- non-leaking PCB transformers and capacitors[47]
- liquid containers where PCB does not exceed 500 ppm, if SPCC plan has been prepared[48]

Specification Storage

- PCB item must be stored so rain does not contact it, i.e., impervious floor, protective walls, and located outside of the 100-year flood plain.[49] The EPA defines this as a building. I believe the above definition includes a drum with drum head or plastic cover stored in an area not designated as a 100-year flood plain.
- PCB items may be stored in a container for up to one year, but disposal must occur within that year.[50] The generator assumes the legal liability if the PCB item is delivered for disposal with less than 90 days remaining within the one-year period.
- Concrete floors used during storage are considered impervious.[51] However, concrete is considered permeable with regard to PCB releases.[52]

---

[42] 40 CFR §761.65 (c) (1)
[43] 40 CFR §761.65 (c) (1)
[44] 40 CFR §761.65 (c) (1) (iv)
[45] 40 CFR §761.65
[46] 40 CFR §761.65 (c) (1) (ii)
[47] 40 CFR §761.65 (c) (1) (i)
[48] 40 CFR §761.65 (c) (1) (iv)
[49] 40 CFR §761.65 (b) (1)
[50] 40 CFR §761.65 (a)
[51] 40 CFR §761.65 (b) (1) (iv)
[52] 40 CFR 761.123

- Liquid PCB may be stored only in the following containers:[53]
  — DOT 17E
  — DOT 6D – overpack drum with 2S or 2L polyethylene liners
- Non-liquid PCB may be stored only in the following DOT designated containers with removal heads:
  — DOT 5
  — DOT 5B
  — DOT 17C

Reclassification of PCB-Containing Equipment

The reuse of PCB transformers placed in storage for reuse is prohibited as of October 1, 1985. However, mineral oil-filled equipment containing PCB may be reclassified by draining, refilling, and/or otherwise servicing.[54] Drained and retrofilled transformers typically reduce from PCB to PCB-contaminated status. Similarly, PCB-contaminated transformers reduce to non-PCB after draining and retrofilling.

Retrofilled transformers must be operated for 90 days at full load [at least 50° C] prior to retesting in order to be reclassified.[55]

The owner of a transformer for reclassification must prepare the following information within 30 days of installation and maintain it indefinitely:[56]

- date of transformer installation
- type of transformer
- PCB concentration at time of installation, if known
- retrofill and reclassification schedule

Retrofilled mineral oil PCB transformers may be used indefinitely.

On November 18, 1993, at 58 FR 60970, the EPA proposed a reclassification schedule based on the original dielectric fluid PCB concentration. As of May 1, 1995, the final rule had not been published. However, on December 6, 1994, at 59 FR 62788, the EPA proposed another revision to 40 CFR Part 761 but has not acted on it yet.

---

[53] 40 CFR §761.65 (c) (6)

[54] 40 CFR §761.303 (a) (2) (i) and 58 FR 60970

[55] 40 CFR §761.30 (a) (2) (iii) and (v)

[56] 40 CFR §761.30 (a) (1)

Decontamination and Salvage of PCB Containers

PCB containers must be decontaminated by triple rinsing with a solvent containing < 50 ppm PCB.[57] PCB solubility in the solvent must be 5% or more by weight. Each rinse must use a volume of solvent equal to 10% or more of the container volume.[58] The solvent must be disposed of under the rules for fluids containing 50-500 ppm PCB [incineration or chemical landfill]. Refer to 40 CFR §761.60 for details on incineration and 40 CFR §761.75 for details regarding landfills.

Solids

Dispose of solids per 40 CFR §761.60(a)(4), i.e., incineration or chemical landfill. Surfaces onto which PCB containing fluids are released may be decontaminated by swabbing with a non-PCB solvent. Cleanliness must be demonstrated by a wipe test.[59] However, concrete surfaces cannot be decontaminated by swabbing. Typically, concrete surfaces must be chipped until the wipe test demonstrates a clean surface [± 2 in.][60]

PCB-Contaminated Equipment

PCB-contaminated equipment [except capacitors] must be drained of all free-flowing liquid prior to disposal.[61] No additional decontamination is required. Disposal of PCB-contaminated equipment must terminate useful life of the equipment, i.e., incineration or smelting.

PCB-contaminated equipment can be used for parts, if the parts are decontaminated to < 50 ppm PCB.[62]

Used Oil Containing PCB

Use of oil containing PCB is rigidly controlled.[63]

---

[57] 40 CFR §761.79 (a)
[58] 40 CFR §761.79 (a)
[59] 40 CFR §761.125
[60] 40 CFR §761.123
[61] 40 CFR §761.60 (b) (3)-(5)
[62] 40 CFR §761.20 (c) (5)
[63] 40 CFR §761.20

Used oil containing any detectable concentration of PCB may not be used as a sealant, coating, or for dust control.[64] Used oil containing 2 ppm but <50 ppm PCB [2 ≥PCB <50 ppm] may not be burned for energy recovery except in:[65]

- incinerators approved for PCB destruction
- high-efficiency boilers whose operators have given notification to the EPA
- permitted *Resource Conservation and Recovery Act* [RCRA] incinerators
- industrial boilers/furnaces subject to 40 CFR Part 266, Subpart E regulations

Used oil containing >50 ppm PCB must be incinerated in an EPA-approved incinerator.[66] Further, PCB must not be recycled to a refinery because of the detrimental affect of the chlorine on the equipment.

PCB Transportation

PCB is a hazardous material as defined by 49 CFR §172.101. Accordingly, transportation of PCB is regulated by DOT's RSPA rules. The rules require preparing a shipping manifest if >1 lb of PCB is transported in a single container on or across public roads or railroads.

Preparation of the manifest is critical, because any hazardous material shipped must be listed before non-hazardous materials or each hazardous material must be listed in a conspicuous color. The description of the hazardous material must be precisely as specified by RSPA.

The current RSPA description is : *RQ, Polychlorinated biphenyls, 9, UN 2315, II.*

The reportable quantity [RQ] on the manifest is not required if <1 lb of PCB is shipped. "RQ" represents the reportable quantity under the *Comprehensive Environmental Response, Compensation and Liability Act* [CERCLA], 40 CFR §302.4, i.e., 1 lb for PCB. "Polychlorinated biphenyls" represents the name of the hazardous substance [material] and "9" represents the hazard class, i.e., "other." "UN 2315" is the United Nations number for PCB. "II" represents "the packing group." Refer to 49 CFR Part 172 [HM-181] for details.

[64] 40 CFR §761.20 (d)
[65] 40 CFR §761.20 (e)
[66] 40 CFR §761.20 (c) (2)

PCB Releases

Releases of PCB [regardless of concentration] ≥1 lb must be reported to the National Response Center [1-800-424-8802] if the PCB enters water, air, or contacts the soil.[67] Releases onto impervious surfaces such as floors or steel surfaces are not reportable. Additionally, releases of PCB ≥50 ppm that contact surface waters, sewer flows, sewage treatment systems, or drinking water supplies must be reported to the EPA Regional Office within 24 hours of discovery.[68]

RECORDS FOR MASTER INVENTORY OF CHEMICALS

Typically, the oil and gas extraction industry has considered itself exempt from TSCA, Section 8(a). This is not true, and there are substantial penalties for failing to comply.

EPA regulations[69] require manufacturers, importers, and processors of chemical substances to report the name and quantity, if ≥ 10,000 lb/yr, of each chemical substance manufactured, imported or processed[70] for the calendar year of 1977.[71] The EPA prepared and published the initial Chemical Substances Inventory in May 1979. This inventory included naturally occurring chemical substances, i.e., crude oil and natural gas.[72] The partial updating of the regulations[73] excluded naturally occurring chemical substances.[74] However, this exclusion does not apply to LPG[75] if production is ≥45 bbls/yr.

The EPA requires an inventory update during the period from August 25 to December 23 starting in 1990 and every four years thereafter.[76] Each person required to submit the initial inventory and update inventory data must maintain records documenting information reported to EPA.[77] Based on the above conditions, only gas plants are subject to updating, at present.

---

[67] 40 CFR §302.6 (a)
[68] 40 CFR §761.120 (a) (2)
[69] 40 CFR Part 710
[70] 40 CFR §§710.3 and 710.5
[71] 40 CFR §710.6
[72] 40 CFR §710.26 (b)
[73] 59 FR 39652, 14 June 1994
[74] 40 CFR §710.4 (d)
[75] 40 CFR §710.4 (b)
[76] 40 CFR ¶710.33 (b)
[77] 40 CFR §710.37

CHEMICAL INVENTORY

E&P operators must verify chemical substances purchased and used for *commercial purposes* are included in the chemical substances inventory under TSCA Section 8(b). These include drilling fluid additives, treating chemicals, maintenance chemicals, and process chemicals. Chemicals not used for commercial purposes [paints, solvents, degreasers] do not require TSCA inventory verification.[78] Typically, verification is effected by requesting the information from the manufacturer.

The manufacturing, importing, processing, or using of chemical substances or mixtures not on the chemical substances inventory violates TSCA and can subject companies or individuals to up to $25,000/day penalties and potential criminal penalties.[79]

Secondary materials burned as fuel, regenerated, wasted, or used as extractants for chemical substances on the inventory are not subject to the chemical substances inventory requirements.[80]

RECORDS OF ALLEGATIONS

TSCA Section 8(c) requires manufacturers, processors, and distributors of chemical substances or mixtures to retain records of allegations of *substantial adverse reactions* to human health or the environment allegedly caused by the substance or mixture.[81] An allegation is a statement without proof or regard for evidence.[82]

Note: Section 8(c) applies to manufacturers, processors, and distributors of chemical substances[83] but does not apply to purchased chemicals not mixed with manufactured products [propane, butane, etc.]. *Further, E&P operations involving only extraction functions, i.e., producing only, are exempt from Section 8(c).*[84] However, E&P facilities operating gas plants, sulfur plants, or CO_2 separation facilities are covered by Section 8(c).

---

[78] 40 CFR §710.4 (d) (60)
[79] 15 USC 2615 (a)
[80] 40 CFR §§710.4 (d) (5) (7)
[81] 40 CFR Part 717
[82] 40 CFR §717.3 (a)
[83] 40 CFR §717.1 (a)
[84] 40 CFR §711.7 (a)

If allegations are received regarding *purchased products*, they need not be retained by the operator. However, they should be given to the supplier of the purchased product.

For non-E&P operators, Section 8(c) requires maintaining the written and signed allegations for 30 years. Verbal allegations must be transcribed by the operator into written form or the alleger must be requested to submit the allegation in a written and signed form.[85]

The allegation must include:

- substance name
- if a mixture, mixture name
- if an article, article name [*Article* is defined as a manufactured item formed into a specific shape having a purpose dependent on the shape and having no change in chemical composition during use.[86]]
- company process or operation name involving the substance
- name of effluent, emission, or other discharge from the site of manufacturing, processing, or distributing the substance causing the adverse reaction "indicating substantial impairment of normal activities or long-lasting or irreversible damage to health or the environment"[87]

The EPA further defines *adverse reaction* by requiring the alleger to record:[88]

- long-lasting or irreversible damage, such as cancer or birth defects
- partial or complete impairment of bodily functions, such as reproductive, neurological, and/or blood disorders
- impairment of normal activities experienced by *all or most* persons exposed at *one time*
- impairment of normal activities experienced each time an individual is exposed

Section 8(c) does not require recording *known reactions*.[89] *Known reactions* are defined as "commonly recognized human health effects

[85] 40 CFR §717.10 (b)
[86] 40 CFR §710.2 (f)
[87] 40 CFR §717.3 (i)
[88] 40 CFR §§717.12 (a) (1) - 717.12 (a) (4)
[89] 40 CFR §717.12 (b)

[caused by] a particular substance or mixture described in... scientific... publications... [and] product Material Safety Data Sheets".[90]

The allegation may be made to the company by any person including an employee, consumer, neighbor, another firm on behalf of an employee, or an organization on behalf of its members.[91]

HEALTH AND SAFETY DATA REPORTING

Section 8(d) requires an operator, who has conducted a health or safety study on certain chemical substances or mixtures, to submit the study to the EPA within 60 days of the substance being added to the master list[92] or within 60 days of proposing to manufacture the substance. However, a number of exceptions apply.

For physical and chemical properties, submit only those involving:[93]

- water solubility
- absorption/desorption on soils
- vapor pressure
- octal/water partition coefficient
- density/specific gravity
- particle size distribution for insoluble solids
- dissociation constant
- degradation by photochemical mechanisms
- degradation by chemical mechanisms
- degradation by biological mechanisms

The EPA must receive a list of initiated studies relating to the listed substances[94] within 30 days of initiation of the study.[95] A copy of the completed study must be sent to the EPA within 30 days of completing the study.[96] Few, if any E&P operators prepare such studies.

---

[90] 40 CFR §717.3 (c) (1)
[91] 40 CFR §717.10 (c)
[92] 40 CFR §716.60 (a) and 716.120
[93] 40 CFR §716.50
[94] 40 CFR §716.35 (a) and 716.120
[95] 40 CFR §716.60 (b) (1)
[96] 40 CFR §716.60 (b) (2)

REPORTING OF SUBSTANTIAL RISK

Any operator manufacturing, processing, or distributing in commerce a chemical substance or mixture who obtains information reasonably supporting the fact the substance or mixture presents a substantial risk to human health and/or the environment is required under Section 8 (a) to report it immediately upon discovery [within 15 working days] to the EPA.[97]

Section 8(e) allows employees to report any suspected hazards directly to the EPA or permits the operator to establish an in-house reporting system.[98] In either case, the operator should place a notice on the bulletin board giving information regarding which alternate is to be used.

The EPA has not promulgated regulations implementing Section 8(e). However, the EPA published an 8(e) Policy Statement on March 16, 1978, at 43 FR 11110. In a *Federal Register* notice dated July 13, 1993,[99] the EPA proposed modifying the TSCA 1978 8(e) policy.[100] As of May 1, 1995, the EPA has not published the final revised policy.

What follows is my interpretation of available compliance procedures.

Direct Contact Procedure

The operator should place a notice on the facility's permanent bulletin board explaining the requirement for employees to report to the EPA any previously *unknown* suspected hazard to human health and/or the environment observed or experienced regarding *a chemical substance or mixture manufactured, processed, or imported by the operator*. The notice should include the telephone number of the EPA Region in which the facility is located. Further, the notice should include a statement to the effect, the employee can be subject to up to a $25,000 fine and one year incarceration for failure to report.[101]

Under this option, employees are authorized to call the EPA directly when reporting a suspect hazard. The operator must instruct his supervisors to encourage reporting to avoid penalties.

---

[97] 15 USC 2607 (e) and 42 FR 45362, 9 September 1977

[98] 42 FR 45362

[99] 58 FR 37735

[100] 43 FR 11110

[101] 15 USC 2615 (b)

Data provided by the employee should approximate that required under Section 8(c) above.

In-House Suspect Hazard Reporting System

Most operators prefer screening the suspect hazard reports by developing an internal reporting system. Again, the operator should place a notice on the permanent bulletin board explaining the requirement for employees to report previously unknown suspected hazards to human health and/or the environment observed or experienced regarding a chemical substance or mixture manufactured, processed, or imported by the operator. Typically, suspect hazard reporting includes:

- A permanent bulletin board notice should be posted stating the operator has an internal reporting system. Each subsidiary is required to report individually. Accordingly, separate notices signed by the facility manager or the health, safety and environmental manager should include instructions for reporting the suspected hazard and include the data required under Section 8(c) above.
- A multipart form should be prepared on which to record the suspected hazard to human health or the environment. The form should include the data required under Section 8(c) above. The foreman, designated alternate, or employee may complete the form. As much specific information as possible should be included on the form, because the information will be evaluated prior to submission to EPA.
- After signing the form, the employee should be given a copy of the competed form. The foreman should retain a copy, and the remaining completed forms should be sent to the superintendent; operations manager; health, safety, and environmental manager, or vice president of operations for evaluation.
- When the completed forms are received at the head office level, the responsible individual reviews the suspected hazard with the company or contract physician or toxicologist.
- If the review team judges the suspected hazard represents a substantial *unknown* hazard, the operator is obligated to forward the information to the EPA immediately, i.e., within 15 working days.

- If the review team judges the suspected hazard does *not* represent a substantial *unknown* hazard, the operator is not obligated to inform the EPA. *The president of the company must concur in the decision not to submit the suspected hazard report.*
- In either event, the operator must inform the employee of the final decision regarding the suspected hazard report.
- The employee has the right to challenge the decision not to report to the EPA. If the final decision [not to report] is judged incorrect, the operator can be subject to civil penalties of up to $25,000 and criminal penalties of up to one year incarceration.[102]

Section 8(e) applies to operator manufactured products, i.e., crude oil, condensate, LPG, sulfur, and any mixtures of purchased produced and operator manufactured products. It is difficult to conceive of any unknown substantial suspected hazard associated with crude oil, condensate, LPG, or sulfur mixtures, if care is exercised in purchasing chemicals on the chemical substance inventory. Nevertheless, operators should inform their employees of their obligation under Section 8(e).

[102] 40 CFR 721.35

6 Federal Insecticide, Fungicide and Rodenticide Act

SUMMARY

The *Federal Insecticide, Fungicide, and Rodenticide Act* [FIFRA][1] only peripherally impacts E&P operations. The act controls application and manufacturing of pesticide as well as use and disposal of controlled containers. Pesticides include herbicides, bactericides, etc. Herbicide weed control conducted by a licensed contractor avoids application concerns. Returning containers to the supplier or triple rinsing and rendering the container useless avoids disposal concerns.

INTRODUCTION

FIFRA controls licensing and application of pesticides and disposal of certain pesticide containers. Pesticide licensing is of no interest to E&P operators because they do not manufacture pesticides, therefore, this discussion will be limited to the application of pesticides and container disposal.

FIFRA basically is a rewrite of the *Insecticide Act of 1910*. FIFRA, as it is known today, was passed in October 1972 as Public Law 96-516. It was amended in 1975, 1978, 1980, and 1988.

---

1 7 USC 136

FIFRA was written to control use, labeling, and registration of pesticides. However, Congress defined pesticides to include any substance intended to:

- prevent, destroy, repel, or mitigate any pest
- regulate, defoliate, or desiccate vegetation

Accordingly, insecticides, fungicides, and rodenticides are covered by FIFRA.

APPLICABILITY

Typically, FIFRA has little impact on E&P operators, because operators do not manufacture pesticides. However, two areas are controlled by FIFRA.

- certification of pesticide applicators[2]
- disposal of pesticide containers[3]

CERTIFICATION OF PESTICIDE APPLICATORS

FIFRA defines pesticides as:

- insecticides
- fungicides
- rodenticides
- biocides
- herbicides[4]

Herbicides frequently are used by onshore operators for weed control at tank batteries. The services of a licensed contract applicator avoid compliance problems.

Biocides are used in open cooling towers and injection well systems to control bacterial growth that destroys cooling tower slates and decreases well injectivity.

Typically, each state imposes different controls on the application of herbicides. Accordingly, no general statement can be made regard-

[2] 40 CFR Part 171

[3] 40 CFR Part 165

[4] 59 FR 42360; 17 Aug. 1994

ing herbicide use. However, I recommend using a licensed contract service for application of herbicides. Certain herbicides are partially controlled, i.e., Roundup, for example.[5] This concentrate, if mixed and applied according to the manufacturer's instructions, may be used by E&P operators in all states without the need for a licensed applicator. Unfortunately, Roundup is expensive.

If the operator desires to use a less expensive herbicide, the operator should seek advice from the state health department or the local agricultural agency. Typically, low concentration, short-lived pesticides can be applied by a non-licensed person. Care must be exercised regarding the assumption that pesticide acceptable in one state is acceptable in another state. Check with the state before using herbicides.

Biocides do not present the same problem, because they typically are used in a non-aerosol system, i.e., metered into the well or cooling tower basin. Accordingly, licensed applicators are not appropriate to biocides and slimicides.

Certification is controlled by the various states, so no specific instructions can be offered. Typically, E&P operators control weeds at tank batteries and well sites by spraying herbicides. Depending on the type and concentrations of the various herbicide components, use of a certified applicator may be required by the state agency. Accordingly, use of a contract service is strongly encouraged. Any operator who plans to apply herbicides using employees should contact the state health or environmental agencies for the state rules governing herbicides.

DISPOSAL OF PESTICIDE CONTAINERS

Reuse of FIFRA-controlled pesticide containers is prohibited unless specific precautions are followed, because residual pesticide can be present in container seams and joints. Accordingly, container disposal is controlled. I recommend returning the container to the supplier. If this is not possible, containers must be triple rinsed[6] with the dilutant used during application, i.e., typically, water. The rinsate must be poured into the active pesticide system, i.e., cooling tower basin or metering tank. The container then must be rendered useless, i.e., cutting the heads from the drum and crushing the resulting cylin-

[5] 59 FR 35714

[6] 40 CFR 165.9 (b)

der. The triple rinsed un-reuseable container may be disposed as a non-hazardous waste in a landfill. Regulations on disposal and prohibitions on reuse of pesticide containers are specific. 40 CFR Part 165.9 provides recommendations for disposal of three classes of containers:

- Group I – combustible containers
- Group II – non-combustible containers
- Group III – combustible and non-combustible containers formerly containing organic mercury, lead, cadmium, and arsenic or inorganic pesticides.

The EPA on February 11, 1994, at 59 FR 6712 propose rules for reusable containers.

Group I Containers

Combustible containers formerly containing organic or metallo-organic pesticides, except mercury, lead, cadmium, or arsenic compounds, may be disposed in a pesticide incinerator, or taken to a specially designated landfill wherein pesticides are buried in a designated cell.[7]

Group II Containers

Non-combustible containers formerly containing organic or metallo-organic pesticides, except organic mercury, lead, cadmium, or arsenic compounds, must be triple rinsed. The container may be returned to the manufacturer or formulator or drum reconditioner for reuse *with the same* chemical class of pesticide as previously contained. Certain DOT rules[8] apply. Accordingly, the best practice is to return the containers to the supplier.

If returning is not practical or is impossible because of damage to the container, the container should be rendered unusable by removing the two heads and crushing the cylinder. This triple rinsed metal can be sold as scrap. *The rinsate must be returned to the active pesticide system for use.*[9]

---

[7] 40 CFR §165.9 (a)

[8] 49 CFR §173.28

[9] 40 CFR §165.9 (b)

Group III Containers

Combustible and non-combustible containers formerly containing organic mercury, lead, cadmium, or arsenic, or inorganic pesticides must be triple rinsed, heads removed, cylinder crushed, and the metal taken to a sanitary landfill. The rinsate must be returned to the active pesticide system for use.[10]

Dangerous Biocides

The most potent FIFRA-controlled biocide found in oil and gas operations is acrolein. Acrolein is maintained under pressure, so a leak results in aerosol formation. Additionally, acrolein is a hazardous substance. Refer to Chapter 8 for details regarding the *Comprehensive Environmental Response, Compensation and Liability Act*. Most operators have discontinued use of acrolein. Formaldehyde is another aggressive biocide. Use should be avoided, because formaldehyde is a carcinogen.

Under no conditions should the containers of these biocides or other FIFRA-controlled chemicals be given away or sold without triple rinsing and rendering them unusable.

[10] 40 CFR §165.9 (c)

7 Resource Conservation and Recovery Act

SUMMARY

The *Resource Conservation and Recovery Act* [RCRA] controls the disposal of all solid waste. The act defines solid waste as non-hazardous and hazardous waste. Severe penalties, both civil and criminal, result from violating the hazardous waste regulations. Additionally, improper disposal of waste can result in large financial liability under the *Comprehensive Environmental Response, Compensation and Liability Act* [Chapter 8]. This is offset by the E&P exemption[1] which classes drilling fluids, produced waters, and associated wastes as exempt from hazardous waste control. E&P operators must understand the limits of the E&P exemption to avoid abusing the provision and losing the exemption.

INTRODUCTION

The RCRA, promulgated October 21, 1976, as 42 USC 6901 et seq., provides for control of solid wastes. The act also was designed to

---

[1] 42 USC 6921 (b) (2) (A)

encourage recycling of materials. A major modification was enacted on July 18, 1984 in the *Hazardous and Solid Waste Amendments*.[2] Basically, solid waste is any discarded material. This means if material is recycled without being discarded or without intent to discard, i.e., placing spent engine oil in the stock tanks for recycling at the refinery, it is not a solid waste. Accordingly, any spent material returned to the manufacturing process of origin is not a solid waste and exempt from RCRA control.

SOLID WASTE

The EPA regulations make the definition of solid waste complex and easily misunderstood. I interpret solid waste to mean any material discarded after completion of its useful life.[3] Only solid waste is controlled by RCRA.[4] Accordingly, any material returned to its original manufacturing process [solvents, engine and gear oils to the stock tanks, reaction vessel catalyst, surplus treating chemical] exempts the material from being classed as a solid waste and from control under RCRA.

RCRA divides solid waste into two classes:

- hazardous wastes[5]
- non-hazardous wastes[6]

Hazardous waste is controlled by the EPA under Subtitle C of RCRA.[7] Typically, states also control hazardous waste, either as a primacy state or under state waste acts. Fortunately, most state rules closely follow the EPA regulations. However, some variations exist in classing certain non-hazardous inert waste as not subject to regulation or, as in the case of California, "all solid waste is a hazardous waste until proven non-hazardous." The reader must contact the state environmental quality department or similar agency to determine the local

---

[2] PL 98-371

[3] 40 CFR §261.2 ((a) (1)

[4] 42 USC 6901 (a)

[5] 40 CFR §261.3 (a) (2)

[6] 40 CFR §261.3 (d)

[7] 42 USC 6921

rules. However, in general, concrete, wood, and stone are considered inert and not subject to analysis.

Hazardous Waste

The EPA divides hazardous waste into two types:

- characteristic wastes[8]
- listed wastes[9]

Typically, only characteristic hazardous wastes are generated in E&P operations.

RCRA Subtitle C imposes "cradle-to-grave" responsibility on the generator of hazardous waste. This means *any hazardous waste* is the responsibility of the generator until the hazardous waste is destroyed or converted to non-hazardous waste. A contract disposer of hazardous waste acts as a *custodian* of that hazardous waste until destruction or conversion of the hazardous waste is completed. *Thus, preventing generation of hazardous waste is much more productive and less expensive than disposing of hazardous waste.*

The operator does not convey the responsibility for hazardous waste previously generated when the property is sold. The generator continues to sustain the responsibility regardless of the number of times the property changes hands. As long as the generator continues to exist, i.e., continues in business even when performing a different type of work, the generator is financially responsible for the hazardous waste.

This concept continues under the *Comprehensive Environmental Response, Compensation, and Liability Act* which holds a generator of hazardous waste financially responsible for a bankrupt hazardous waste site.[10] Refer to Chapter 8 for details.

---

[8] 40 CFR §261.3 (a) (2) (i)

[9] 40 CFR §261.3 (a) (2) (iii)

[10] Parrish, P.G. "Comprehensive Environmental Response, Compensation, and Liability Act of 1980 as amended [CERCLA] — Corporate Liability: Successor, Parent, Stockholder, and Officers," Presented at Environmental Law in the Oil and Gas Industry, University of Oklahoma, Oct. 22-23, 1990.

Characteristic Waste

Four waste characteristics are defined by the regulations:

- ignitability[11] [I]
- corrosivity[12] [C]
- reactivity[13] [R]
- toxicity[14] [T]

Typically, E&P hazardous wastes are ignitable or corrosive.

Ignitability Characteristic

Ignitable wastes can be either liquids or solids. Liquid waste exhibiting the characteristic of ignitability has a closed cup flash point [ASTM D-93-79 or 80][15] of less than 140° F [FP <140° F][16] Solid waste exhibiting the characteristic of ignitability can be ignited at standard conditions and will sustain a flame without assistance from an ignition source.[17]

No ignitable solid waste is found in E&P operations. On the other hand, the majority of oil and gas treating chemicals are ignitable liquid hazardous wastes, *if wasted*. Thus, do not dispose of treating chemicals [methanol, dehydration chemicals]. Instead, use the surplus chemical for its intended purpose or return to the supplier.

A word of caution is necessary at this point. Use "oil treating chemicals" for treating oil and "water treating chemicals" for treating water. Misusing a chemical can result in accusation by the EPA or the state of "disposing" of the chemical. This can make the oil or water treating facility a waste treatment facility. Accordingly, the lease operator should maintain a contemporaneous record of chemical usage so the operator can document the chemical was used properly. Disposing of ignitable wastes is very expensive, typically by incineration.

---

[11] 40 CFR §261.21
[12] 40 CFR §261.22
[13] 40 CFR §261.23
[14] 40 CFR §261.24
[15] 40 CFR §260.11
[16] 40 CFR §261.21 (a) (1)
[17] 40 CFR §261.21 (a) (2)

Many liquid hydrocarbons are not ignitable wastes. Lubricating oil and gear oils have flash points >140° F. These oils can be disposed of as non-hazardous waste. However, recycling these oils via the crude oil tanks eliminates transportation costs associated with re-refining and increases net operating income, because the spent oils are sold as crude at a higher price than re-refiners pay. Also, spent non-chlorinated solvents can be recycled via the crude oil tank. In some cases, mixing lubricating oil with condensate discolors the condensate, rendering the mixture non-marketable. In this case, transport the spent lubricating oil to an oil lease. Admittedly, royalty must be paid on the lubricating oil. However, the operator makes a net profit by eliminating a waste.

Care must be taken not to recycle chlorinated hydrocarbons to the refinery via the crude oil tanks, because high temperature chlorine attacks the refinery piping and vessels. Typically, only two chlorinated hydrocarbons are observed in E&P operations:

- trichloroethane
- dielectric fluid containing PCB

Trichloroethane is very seldom used outside of gas plants and in these only sparingly. On the other hand, some large old leases may have operator-owned transformers containing PCB. Disposal of PCB is controlled by TSCA [Chapter 5]. Recycling hydrocarbons containing PCB via the crude oil tanks is deleterious to the refinery and violates TSCA.

Corrosivity Characteristic

Any aqueous liquid waste demonstrating pH less than or equal to 2.0 units or equal to or greater than 12.5 [2.0 ≤ pH ≤ 12.5] is classed a corrosive waste.[18] Also, liquid wastes causing a loss of steel [SAE

[18] 40 CFR §261.22 (a) (1)

1020] at the rate greater than ¼ inch per year at a test temperature of 130° F is a corrosive waste.[19]

Typically, E&P operators use hydrochloric acid as a stimulation agent in carbonate formations. Also, some gas plants use sulfuric acid for pH control of cooling tower water. If wasted, these acids would be corrosive wastes. However, unused acids returned to the supplier for use are not solid wastes. Further, acids returned during backflow of production are statutorily exempted.[20] Accordingly, corrosive wastes seldom are observed in E&P operations. Typically, spent acids are considered produced water and injected into disposal wells. Cooling tower blowdown may be injected into a disposal well, if the gas plant is an in-field plant. Refer to Chapter 4 for limitations.

Reactivity Characteristic

Any solid waste reacting violently when mixed with water or other chemicals; generating toxic gases; generating excessive heat or exploding when subjected to heat is a reactive waste.[21] No quantitative test of reactivity exists. The agency chose to rely on a descriptive, prose definition of reactivity because the available tests for measuring the variegated class of effects embraced by the reactivity definition suffer from a number of deficiencies. To the best of my knowledge, the only possible reactive waste associated with E&P operations is concentrated sulfuric acid. However, acids are seldom wasted, because they are returned to the supplier if no longer needed.

Toxicity Characteristic

Waste toxicity is derived from heavy metals, pesticides, and organics, Table 7–1.[22] The EPA-approved test method is known as the toxicity characteristic leaching procedure [TCLP].[23] Heavy metals are

[19] 40 CFR §261.22 (a) (2)
[20] 42 USC 6921 (b) (2) (A)
[21] 40 CFR §261.23 (a) (1-8)
[22] 40 CFR §261.24 (b)
[23] 40 CFR Part 261, Appendix II

observed in produced water and drilling fluids.[24] However, these solid wastes are exempted by statute from being classed as hazardous under RCRA.[25] Few other wasted materials observed in E&P operations contain heavy metals. Some heavy metal liquid paints may be found in offshore operations. However, these materials should be used as prime coats. Before discarding paint brushes, they should be allowed to dry without cleaning, because the TCLP will not cause dried paint to release sufficient heavy metals to become a hazardous waste.

The TCLP[26] basically requires waste be ground, placed in a glass column, through which a weak acid is flowed. The resulting leachate is analyzed for the toxic characteristic chemicals. The test is sensitive and the regulatory levels low, Table 7–1.

The toxic chemical of greatest concern in E&P operations is benzene [0.5 mg/l regulated level], because all oils and condensates contain benzene. Storm water contacting a fresh release of crude oil or condensate may pick up benzene. Liquid wastes [storm waters] are considered leachates. The liquid waste is diluted 20:1 before analysis or the liquid waste is analyzed and resulting concentration divided by 20. Therefore, liquid wastes, such as rain water, contacting fresh oil or condensate seldom become toxic wastes. Nevertheless, avoid hydrocarbon releases, because sufficient benzene can be picked-up from very fresh spills to make storm water toxic, thus, invoking the storm water permit provisions of the *Clean Water Act*, Chapter 2. This brings into sharp focus the need to keep tank battery dike drains closed and sealed.

---

[24] Deuel, L.E. Jr. and G.H. Holliday, *Soil Remediation for the Petroleum Extraction Industry* (Tulsa: PennWell, 1994).

[25] 42 USC 6921 (b) (2) (A)

[26] 40 CFR Part 261, Appendix II

Table 7–1 Maximum concentration of contaminants for the toxicity characteristic

EPA HW No.[27]	Contaminant	CAS No.[28]	Regulatory Level(mg/L)
D004	Arsenic	7440-38-2	5.0
D005	Barium	7440-39-3	100.0
D018	Benzene	71-43-2	0.5
D006	Cadmium	7440-43-9	1.0
D019	Carbon tetrachloride	56-23-5	0.5[29]
D020	Chlordane	57-74-9	0.03
D021	Chlorobenzene	108-90-7	100.0
D022	Chloroform	67-66-3	6.0
D007	Chromium	7440-47-3	5.0
D023	o-Cresol	95-48-7	200.0[29]
D024	m-Cresol	108-39-4	200.0[29]
D025	p-Cresol	106-44-5	200.0[29]
D026	Cresol		200.0[29]
D016	2,4-D	94-75-7	10.0
D027	1,4- Dichlorobenzene	106-46-7	7.5 [29]
D028	1,2- Dichloroethane	107-06-2	0.5 [29]
D029	1,1- Dichloroethylene	75-35-4	0.7
D030	2,4- Dinitrotoluene	121-14-2	0.13[30]
DO1	2Endrin	72-20-8	0.02
D031	Heptachlor (and its epoxide)	76-44-8	0.008[29]
D032	Hexachlorobenzene	118-74-1	0.13[30]
D033	Hexachlorobutadiene	87-68-3	0.5
D034	Hexachloroethane	67-72-1	3.0
D008	Lead	7439-92-1	5.0
D013	Lindane	58-89-9	0.4
D009	Mercurt	7439-97-6	0.2
D014	Methoxychlor	72-43-5	10.0
D035	Methyl ethyl ketone	78-93-3	200.0
D036	Nitrobenzene	98-95-3	2.0
D037	Pentrachloropheno	l2987-86-5	100.0
D038	Pyridine	110-86-1	5.0[30]
D010	Selenium	7782-49-2	1.0
D011	Silver	7440-22-4	5.0
D039	Tetrachloroethylene	127-18-4	0.7
D015	Toxaphene	8001-35-2	0.5
D040	Trichloroethylene	79-01-6	0.5
D041	2,4,5- Trichlorophenol	95-95-4	400.0
D042	2,4,6- Trichlorophenol	88-06-2	2.0
D017	2,4,5-TP(Silvex)	93-72-1	1.0
D043	Vinyl chloride	75-01-4	0.2

[27] Hazardous waste number

[28] Chemical abstracts service number

[29] If o-, m-, and p-Cresol concentrations cannot be differentiated, the total cresol (DO26) concentration is used. The regulatory level of total cresol is 200 mg/l.

[30] Quantitation limit is greater than the calculated regulatory level. The quantitation limit therefore becomes the regulatory level.

Typically, heavy metals, pesticides, and organics do not cause E&P wastes to be toxic.[31]

Listed Wastes

The EPA developed several categories of listed wastes: [32]

F- wastes	from non-specific sources [33]
K -wastes	from specific sources by industries [34]
P - wastes	discarded commercial chemical products, manufactured chemical intermediates, or off-specification commercial chemical products, or manufacturing chemical intermediates – classed as acute hazardous waste [H] [35]
U-wastes	discarded commercial chemical products, manufacturing chemicals, intermediates, or off-specification commercial chemical products – typically [but not exclusively] classed as toxic waste [T][36]

Typically, E&P operations do not generate listed waste. However, a few E&P chemicals can be listed waste, i.e.:

• acrolein	POO3	[H]
• carbon disulfide	PO22	[H]
• formaldehyde	U122	[T]
• 1,1,1 - trichloroethane	U226	[T]
• tetrahydrofuran	U213	[T]
• methyl ethyl ketone [MEK]	U159	[I,T]
• pyridine	U196	[H]

Generally, with the exception of carbon disulfide, 1,1,1-trichloroethane, MEK, and pyridine, the use of chemicals resulting in

---

[31] Duele, L.E. and G.W. Holliday: *Soil Remediation for the Petroleum Extraction Industry* (Tulsa: PennWell, 1994).

[32] 40 CFR §§261.30-261.33

[33] 40 CFR §261.31

[34] 40 CFR §261.32

[35] 40 CFR §261.33 (e)

[36] 40 CFR §261.33 (f)

the generation of hazardous waste should be eliminated. First, acrolein is a very toxic chemical that can be absorbed through the skin. Second, formaldehyde is a known carcinogen. Accordingly, both chemicals should be eliminated and substitutes used.

Further, avoid disposing of any hazardous wastes by using the substances for their intended purposes.

Waste Classification Determination

The EPA allows three methods for determining whether a solid waste is hazardous:

- characteristic test failure
- located as a listed waste
- knowledge of the waste[37]

Failure of the waste to pass the ignitibility, corrosivity, reactivity, or toxicity tests[38] or inclusion of the waste within any of the lists[39] requires classing the waste as hazardous.

The EPA allows generators the option of not testing waste, if the generator has knowledge of the process or waste sufficient to make a reasoned decision.[40] This "knowledge" can be based on previous testing, raw material information, or on process understanding. In the event the generator determines the waste to be hazardous, the following certification must be signed:[41]

> I certify under penalty of law that I personally have examined and am familiar with the waste through analysis and testing or through knowledge of the waste to support this certification that the waste complies with the treatment standards specified in 40 CFR Part 268 Subpart D and all applicable prohibitions set forth in 40 CFR 268.32 or RCRA Section 3004(d). I believe that the information I submitted is true, accurate and complete. I am aware that there are significant penalties for submitting a false certification, including the possibility of a fine and imprisonment.

---

[37] 40 CFR §268.7

[38] 40 CFR §261.20

[39] 40 CFR §§261.30-261.33

[40] 40 CFR §268.7 (a)

[41] 40 CFR §268.7 (a) (2) (ii)

In the event the waste is determined to be non-hazardous on the basis of "knowledge," I strongly recommend a "note to the file" be generated justifying the determination.

Non-Hazardous Waste

Any *solid waste* which is not a hazardous waste is a non-hazardous waste. Currently, no non-hazardous federal regulations have been promulgated. On the other hand, states promulgate regulations controlling analysis and disposal of non-hazardous waste.

Many states prohibit disposal of industrial non-hazardous waste in city and county landfills. Accordingly, industrial non-hazardous waste is handled best through a refuse service. However, no commercial refuse service should be used unless the disposal site is visited and judged to be properly operated, i.e., working face covered with soil daily, paper screens in place to avoid blowing paper. Most importantly, there must be a system in place to avoid accepting hazardous waste at the non-hazardous waste site. A visit by the E&P operator foreman or supervisor is essential to avoid using a poorly operated landfill which at a later date may be judged a hazardous waste site.

Waste Mixing

Wastes must be segregated into hazardous and non-hazardous wastes, except for conditionally exempted small quantity generators as defined below.[42] Waste segregation is of utmost importance when the operator is a large quantity generator.

The EPA promulgated complex and stringent rules regarding mixing of hazardous and non-hazardous waste. Most E&P operators generate, at worst, only small amounts of hazardous waste. Accordingly, the mixture rules are of passing interest.

For simplicity, mixing of a hazardous waste with a non-hazardous waste makes the mixture a hazardous waste. While this interpretation is not precise, it serves as a viable guide. However, for completeness, the mixture rules are interpreted.

---

[42] 40 CFR §261.5

Characteristic Waste Mixing

Mixing of a characteristically hazardous waste with non-hazardous waste where the mixture retains the original characteristic makes the mixture a hazardous waste.[43] If, for some strange reason, the mixture exhibits a new hazardous waste characteristic, the mixture is a hazardous waste by definition.[44]

Listed Waste Mixing

A mixture of a listed hazardous waste with a non-hazardous waste renders the mixture a hazardous waste, except if the hazardous waste is listed *only* because of a characteristic, i.e., ignitability, toxicity, corrosivity or reactivity,[45] of a hazardous waste and the resulting mixture no longer exhibits the characteristic causing the waste to be listed.

Waste Handling Priorities

To avoid contravening the hazardous waste regulations, prioritize waste handling options, Table 7–2. Basically, the priorities of waste generation run from "avoid generating waste" to "dispose of waste in an inspected commercial site."

Table 7–2 Waste handling priorities

A. Eliminate waste production
B. Minimize waste production
C. Recycle waste onsite
D. Recycle at another company facility
E. Recycle waste at company approved contract facility
F. Dispose waste on site
G. Dispose waste at another company facility
H. Dispose waste at a company approved contract facility

---

[43] 40 CFR 261.3 (a) (2) (i)
[44] 40 CFR 261.3 (a) (2) (i)
[45] 40 CFR §261.3 (a) (2) (iii)

Waste Storage

Non-Hazardous Waste

Non-hazardous waste should be stored in metal containers [open-topped drums with lids]. The contents should be transferred to a commercial dumpster at least weekly. Typically, commercial dumpsters are located at the operator office or central "doghouse." In some cases, commercial service is not available. In this case, the operator should provide a closed-top trailer that can be taken to a commercial, city, or county landfill. *None of these landfills will accept waste liquids*. Accordingly, if the spent liquid cannot be recycled, the liquid must be stabilized by mixing the liquid with sawdust, pozzolan, cement, or other absorbent materials.

Hazardous Waste

If hazardous waste is generated >220 lb/month [see exempted waste below], the hazardous waste regulations apply to the waste.[46] Storage and shipment of hazardous waste must occur in DOT approved containers.[47]

At the time storage of hazardous waste begins, the container must be dated.[48] Waste may be stored up to 90 days,[49] provided the waste is placed in a container:[50]

- in good condition
- compatible with the waste
- closed except when adding or removing waste
- inspected at least weekly

Additionally, the container must be placed on an undrained drip pad and the following records maintained:[51]

- description of procedure assuring waste is removed every 90 days
- document each waste removal including date and time of removal

[46] 40 CFR §261.5 (g)
[47] 40 CFR §263.10
[48] 40 CFR §262.34 (a) (2)
[49] 40 CFR §262.34 (b)
[50] 40 CFR §§265.170-265.177
[51] 40 CFR §262.34 (a) (2)

Further, each hazardous waste container must be labeled "Hazardous Waste."

Notwithstanding the above requirements, a generator may accumulate as much as 55 gal of hazardous waste or 1 qt of acutely hazardous waste [P-waste] at or near the point of generation under the control of the operator[52] provided:

- container labeled "Hazardous Waste"
- complies with the above container rules for hazardous waste

However, within the 90 days, the waste must be shipped and disposed.[53] Containers ready for shipment must be labeled as follows:

- proper DOT shipping name[54]
- numeric hazard class[55]
- DOT identification number [UN number][56]
- DOT required packaging specification, i.e., DOT 17E, 17C etc.[57] [After October 1, 1996, UN specification packaging will be required. Refer to 59 FR 36488, July 18, 1994, for proper shipping names of contaminated solids.]
- shipper's name and address[58]
- disposal site waste stream approval number
- hazardous waste label[59]
- reportable quantity, if container holds more than the reportable quantity of a hazardous substance.[60]

Any shipment of a hazardous waste across U.S. territorial seas must be accompanied with a Coast Guard Dangerous Cargo Manifest.[61] Shipment of hazardous waste from onshore sites must be ac-

[52] 40 CFR 262.34 (c) (1)
[53] 40 CFR §262.34
[54] 49 CFR §172.101
[55] 40 CFR §§261.21-261.33
[56] 49 CFR §172.101 (e)
[57] 49 CFR §172.101 (c) (10) (A)
[58] 49 CFR §172.205
[59] Mark as follows: HAZARDOUS WASTE – Federal Law Prohibits Improper Disposal. If found, contact the nearest police or public safety authority or the U.S. Environmental Protection Agency.
Generator's Name and Address____________________
Manifest Document Number____________________
40 CFR §§262.31 & 262.32
[60] 40 CFR §302.4
[61] 33 CFR §148.10 and 49 CFR §176.30

companied by a Hazardous Waste Manifest.[62] The Hazardous Waste Manifest is a multipart form designed to assure chain of custody of the waste.

The generator completes the form describing the amount, type, and destination of the waste.[63] Also, the generator provides his identification number. The transporter provides the transport vehicle identification and signs as having received the waste identified on the manifest.

The operator takes the first two copies and retains one copy and sends the other copy to the state [primacy states] or the EPA [nonprimacy states]. The remaining copies go with the waste.[64] If the waste goes to storage, the storer signs as having received the waste identified on the manifest from the vehicle identified on the manifest. Similarly, the disposer signs the manifest when the waste is received. A copy of the manifest is retained by each organization handling the waste. The disposer retains a copy and sends one copy to the state or the EPA.

If the generator of >1,000 kg/month [2,200 lb] does not receive the signed copy of the manifest from the disposer within 35 days of the waste acceptance by the transporter, the generator must contact the transporter or disposal facility operator to determine the status of the waste.[65] The generator must send to the EPA regional administrator an Exception Report if the signed copy of the manifest is not received within 45 days of shipment.[66] The Exception Report must include:

- copy of the Hazardous Waste Manifest
- signed statement by the generator of the attempts to locate the waste and the results of these efforts

If the generator of >100 kg/month [220 lb] but <1,000 kg/month [2,200 lb] fails to receive a signed copy of the waste manifest within 60 days of waste shipment, the generator must forward to the regional administrator a copy of the manifest and an indication the signed manifest has not been received.[67] The generator is obligated to compare the original copy with the final copy. If the manifests match, both copies are filed and retained for three years. If the manifest copies do not match, i.e., sent to the wrong storer or generator, different amounts

[62] 40 CFR §262.20
[63] Appendix to 40 CFR Part 262
[64] 40 CFR §262.23 (b)
[65] 40 CFR §262.42 (a) (1)
[66] 40 CFR §262.42 (a) (2)
[67] 40 CFR §262.42 (b)

received than sent, etc., the disposer within 15 days must notify the state or the EPA.[68]

Reports and Recordkeeping

Generators shipping hazardous waste offsite to a treatment, storage, or disposal [TSD] site within the United States must prepare and submit a biennial report to the regional administrator by March 1, of each even numbered year. The report must be submitted on EPA Form 8700-134 and cover hazardous waste generation activities during the previous year.[69] The report must include:

- generator identification number,[70] name, and address
- calendar year covered by report
- TSD name, address, and identification number of U.S. site used during the reporting year
- description of each hazardous waste shipped to U.S. TSD facility during reporting year
- description of activities taken to reduce volume and toxicity of waste generated
- description of achieved reduction of volume and toxicity for reporting year compared with previous years
- signed certificate of the correctness of the report

All hazardous waste reports must be retained for at least three years from the time the waste was shipped.[71]

Exempted Wastes

Several exemptions are provided under RCRA regulations:

- conditionally exempted small quantity generator
- small quantity generator
- E&P exemption

I recommend using the exemptions when they apply.

---

68 40 CFR §264.73
69 40 CFR §262.41 (a)
70 40 CFR §262.12
71 40 CFR §262.40 (a)

Conditionally Exempted Small Quantity Generator

Any person [individual or corporation] generating ≤100 kg/month [220 lb] of a characteristic or listed hazardous waste [except P-waste] per month and not storing ≥1,000 kg/month is exempt from RCRA hazardous waste regulations.[72] This allows any person to dispose of household cleaners, heavy metal paints, ignitable solvents, etc. without the control and manifest procedure discussed above. Typically, E&P operations generate very small quantities of hazardous wastes. Accordingly, operator should use the exemption, but judiciously.

For example, heavy metal paints may be disposed of in a non-hazardous landfill. However, rather than disposing of the liquid paint, boards or cardboard should be painted or the liquid paint allowed to harden before disposal. This procedure eliminates the availability of the heavy metal to the environment. Similarly, other liquids should be absorbed by sawdust, cement, clay, or similar solids. Reactive wastes, i.e., acids, should be neutralized. If neutralization is not possible, the waste should be handled as a hazardous waste.

Small Quantity Generator

Any person generating >100 kg/month [220 lb] but <1,000 kg/month [2,200 lb] is exempted from most provisions of RCRA hazardous waste controls.[73] However, a small quantity generator must obtain an EPA identification number[74] and manifest the hazardous waste shipments.[75] Further:

- the quantity of waste accumulated may not exceed 6,000 kg[76]
- storage of ignitable or reactive waste must be at least 50 ft from property line[77]
- store waste on a drip pan[78]
- storage containers are labeled "Hazardous Waste"
- comply with preparedness and prevention provisions of 40 CFR §§265.30 - 265.37

---

[72] 40 CFR §261.5 (b)
[73] 40 CFR §260.10
[74] 40 CFR §262.12
[75] Appendix to 40 CFR Part 262
[76] 40 CFR §262.34 (d) (1)
[77] 40 CFR §265.176
[78] 40 CFR §262.34 (a) (2)

Additionally small quantity generators must provide:[79]

- at least one employee on call [emergency coordinator]
- name and telephone number of emergency coordinator
- telephone number of fire department
- training for employees in handling of wastes in emergency situations
- in the event of a fire, explosion, or release, provide report of incident

A small quantity generator may store hazardous waste for up to 180 days without a permit, if the above conditions are met.[80] These wastes can be stored for 270 days, if the waste must be transported 200 miles or more.[81]

Also, a small quantity generator can accumulate more than 6,000 kg for more than 180 days or 270 days [if transported 200 miles or more], if the generator complies with 40 CFR Parts 264 and 265 and obtains a permit under 40 CFR Part 270.[82]

E&P Exemption

The most far reaching exemption available to the petroleum extraction industry is the E&P exemption.[83] This exemption classes the following wastes as statutorily non-hazardous:

- drilling fluids, including completion fluids and workover fluids
- produced water
- associated wastes

Drilling fluids and produced water are easily identified. On the other hand, the term *associated wastes* means different things to different people. In my opinion, associated wastes refer to those materials that come up the casing or tubing and that are introduced into the gas and oil treating system in the process of drilling, recompleting, and producing the well, Table 7–3. Accordingly, acids, scale inhibitors, biocides, treating chemicals, produced sand, iron sponge, sulfa-check

[79] 40 CFR §262.34 (a) (5)
[80] 40 CFR §262.34 (e)
[81] 40 CFR §262.34 (e)
[82] 40 CFR §262.34 (f)
[83] 42 USC 6921 (b) (2) (A)

solutions, etc. are classed as non-hazardous waste. These E&P wastes are intrinsic to drilling and producing operations. On the other hand, cleaning solvents, paints, lubricants, etc. are non-exempt wastes and must be considered hazardous if the waste exhibits a characteristic or is a listed waste, Table 7–4.

Preserving the E&P exemption is important to E&P operators, because it allows onsite disposal of drilling fluids and produced water without need for testing.

Table 7–3 Associated wastes

- Rigwash
- Drilling fluids and cuttings from offshore operations disposed of onshore
- Well completion, treatment and stimulation fluids
- Basic sediment and water and other tank bottoms from storage facilities that hold product and exempt waste
- Accumulated materials such as hydrocarbons, solids, sand and emulsion from production separators, fluid treating vessels and production impoundments
- Pit sludges and contaminated bottoms from storage or disposal of exempt wastes
- Workover wastes
- Gas plant dehydration wastes, including glycol-based compounds, glycol filters, filter media, backwash, and molecular sieves
- Gas plant sweetening wastes for sulfur removal, including amine, amine filters, amine filter media, backwash, precipitated amine sludge, iron sponge and hydrogen sulfide scrubber liquid and sludge
- Spent filters, filter media, and backwash (assuming the filter itself is not hazardous and the residue in it is from an exempt waste stream)
- Packing fluids
- Produced sand

  Pipe scale, hydrocarbon solids, hydrates, and other deposits removed from piping and equipment prior to transportation
- Hydrocarbon bearing soil
- Pigging wastes from gathering lines
- Wastes from subsurface gas storage and retrieval, except for the listed non-exempt wastes
- Constituents removed from produced water before it is injected or otherwise disposed of
- Liquid hydrocarbons removed from the production stream but not from oil refining
- Gases removed from the production stream, such as hydrogen sulfide, and carbon dioxide and volatilized hydrocarbons
- Materials ejected from a producing well during the process known as blowdown

— Waste crude oil from primary field operations and production
— Light organics volatilized from exempt wastes in reserve pits or impoundments or production equipment
— Excess cement slurries and cement cuttings
— Sulfur contaminated soil or sulfur waste from sulfur recovery units.
— Gas plant sweetening unit catalyst
— Produced water contaminated soil
— Wastes from the reclamation of tank bottoms and emulsions when generated at a production location
— Production facility sweetening and dehydration wastes
— Production line hydrotest/preserving fluids utilizing produced water
— Iron sulfide

Table 7–4 Non-exempt wastes

— Unused fracturing fluids or acids
— Gas plant cooling tower cleaning wastes
— Painting wastes
— Oil and gas service company wastes, such as empty drums, drum reinstate, vacuum truck rinsate, sandblast media, painting wastes, spent solvents, spilled chemicals, and water wastes
— Vacuum truck and drum rinsate from trucks and drums transporting or containing non-exempt waste
— Liquid and solid wastes generated by crude oil and tank bottom reclaimers
— Lubrication oils
— Waste compressor oil, filters, and blowdown
— Hydraulic fluids
— Waste solvents
— Waste in transportation pipeline related pits
— Waste solvents
— Caustic or acid cleaners
— Boiler cleaning wastes
— Boiler refractory bricks
— Incinerator ash
— Laboratory wastes
— Sanitary wastes

8 Comprehensive Environmental Response, Compensation, and Liability Act

SUMMARY

The Comprehensive Environmental Response, Compensation, and Liability Act [CERCLA] has the potential to impose costly liability on E&P operators. Colloquially, CERCLA is known as Superfund because the act imposes on chemical and refining operations a fee designed to remediate orphaned hazardous waste sites. Improper *disposal of E&P hazardous or non-hazardous waste* can result in imposition of joint and several liabilities, meaning any user of a Superfund site is individually liable for total site remediation costs. Typically, the more financially capable organization pays all or the largest share of the remediation cost.

CERCLA also requires immediate notification of the release of a reportable quantity of a hazardous substance to air, water, or soil.

INTRODUCTION

CERCLA[1] was enacted into law July 18, 1982, as a means of cleaning up orphaned waste sites left by now non-existent companies. The act imposes a liability on disposers and transporters of hazardous waste for removal and remediation costs at orphaned hazardous waste

---

[1] 42 USC 9601

sites.[2] Superfund waste sites remediated by the EPA have cost an average of $25 million each. Accordingly, most sites are closed by the "potentially responsible parties" [PRP].

CERCLA provides for non-hazardous waste sites becoming Superfund sites if the waste imposes an "imminent endangerment to human health and the environment".[3] Accordingly, operators must avoid improper disposal of drilling fluids and produced waters, because the E&P exemption under RCRA does not extend into CERCLA.[4]

Finally, CERCLA requires reporting of releases at or above the specified reportable quantity of hazardous substances[5] immediately upon knowledge of the release occurrence.

WASTE SITE REMEDIATION

CERCLA raises to a new high the importance of waste disposal. Disposal of non-hazardous waste in a hazardous waste site results in the generator of the non-hazardous waste becoming a user of a hazardous waste site and subject to all CERCLA Superfund provisions. *Operators must know where their waste goes*. The days of calling a waste hauler, loading the truck, and feeling secure of no future liability are gone.

If waste goes to a hazardous waste site that at a later time goes into bankruptcy, the generator retains responsibility for the waste, i.e., cradle-to-grave responsibility. The disposer is nearly a custodian of the waste. Accordingly, if a hazardous waste must be used, select a site operated by a financially secure company. Visit the site, observe the operations, look at the records, and know how the waste is handled. Above all, use a hazardous waste site used by a very large and financially sound generator. If a problem develops at a later date, small operators know their interests will be protected by the large corporation in the process of looking out for its own interests.

Typically, E&P operators become involved in Superfund sites through an innocent action of disposing of non-hazardous waste without specifying the site. The EPA identifies generators using a Superfund

---

[2] 42 USC 9607

[3] 42 USC 9606 (a)

[4] Radford. N.D. "An Overview of CERCLA Liability and Its Application to Oil and Gas Industry," presented at Environmental Law in the Oil and Gas Industry, University of Oklahoma, Oct. 22-23, 1990.

[5] 40 CFR §302.4

site via trucking records].[6] These generators are known as "potentially responsible parties" [PRP]. Once a list of PRPs is developed, each PRPs receives a "demand letter" requesting all records relating to the suspected use of the site. The letter typically provides 10 days during which to respond. Often, such records are not immediately available. Accordingly, it is advisable for the operator to respond within the prescribed time and request additional time to develop the requested information. Unless the PRP can demonstrate with evidence the site was *not* used, the operator remains a PRP.

Typically, a negotiating team of attorneys and a technical team of engineers are formed. Through a set of meetings, a remediation plan is developed by PRPs and accepted by the EPA or the state. Proportionate funding, also, is established.

Care must be taken while developing the remediation plan, because simply moving the waste to another site means the PRPs have only created another hazardous waste site that potentially could develop into another Superfund site.

The vast majority of remediations of Superfund sites to date have been associated with ground water pump and treat technology that results in very small quantities of hazardous waste for future disposal.

Unfortunately, Superfund status is not reserved only for orphaned hazardous waste sites. CERCLA Section 9606(a) provides for classing any facility a Superfund site if the waste poses a substantial endangerment to human health or the environment. Currently, four E&P commercial disposal sites are judged Superfund sites by EPA Region VI; 3 in Louisiana[7] and 1 in New Mexico.[8] Generators of waste disposed at these sites are obligated to fund remediation.

PETROLEUM EXCLUSION

CERCLA excludes "petroleum, including crude oil or any fraction thereof, which is not otherwise specifically listed or designated as a hazardous substance [under other environmental statutes] but the term does not include natural gas, natural gas liquids, liquefied natural gas, or synthetic gas usable for fuel..."[9] This exemption is termed the petroleum exclusion.

---

[6] 54 FR 41021; 54 FR 13296; 54 FR 13296; 55 FR 35509

[7] 54 FR 13296 and 41021

[8] 55 FR 35509

[9] 42 USC 9601 (14)

Note: the petroleum exclusion does not include wastes such as drilling fluids and produced waters. Accordingly, the CERCLA petroleum exclusion differs from the RCRA E&P exemption.[10]

The petroleum exclusion can be lost by mixing petroleum wastes with hazardous wastes before disposal.[11]

The statute also includes protection for E&P operators in the form of "federally permitted releases" that include "any injection of fluid authorized under Federal underground injection control programs or State programs submitted for Federal approval [and not disapproved by the Administrator of the Environmental Protection Agency."[12]

MEANS OF LIMITING CERCLA LIABILITY

Careful selection of waste transporters and disposers can limit CERCLA liability:

- Initial price may be the least important factor.
- Determine whether the transporter and disposer have the requisite permits.
- Investigate contractors for compliance with permits.
- Include in contracts with the transporter and disposer that their operations are in compliance with statutes, rules, and permits, and include a strong indemnity provision.
- Make sure the transporter maintains adequate insurance covering sudden and non-sudden occurrences.
- Ascertain the ability of the transporter and disposer to fulfill the indemnity obligations.
- Periodically audit transporter and disposer operations for compliance.

HAZARDOUS SUBSTANCES

The EPA designates a group of chemical substances as hazardous substances.[13] A limited number of these hazardous substances are used in E&P operations, Table 8–1. [Refer to Appendix 4 for full list of

---

[10] 42 USC 6921 (b) (2) (A)

[11] State of Washington v. Time Oil Company, 687F Supp. 529,532 (W.D. Wash.), 1988

[12] 42 USC 9601 (10)

[13] 40 CFR §302.4

hazardous substances.] Further, the EPA established reportable quantities [RQ] for each hazardous substance.[14] Depending on the hazardousness of the substance, the RQ varies from 1 lb to 5000 lb. The EPA proposed revised RQs for many hazardous substances October 22, 1993.[15] However, as of May 1995 the new RQs were not adopted.

The rule requires reporting the release of an RQ or more immediately upon discovery.[16] Further, the report is required if the RQ release occurs over a time period of 24 hours or less. The EPA mandates the report be made to the National Response Center.[17] Release means the substance enters the environment [air, water, or soil]. Accordingly, the spilling of a large quantity of low volatility liquid onto a concrete surface would not represent an RQ release because the liquid did not enter the environment. On the other hand, if a large quantity of chlorine is released to the air, it would be classed as an RQ release.

Most oil and gas treating chemicals contain hazardous substances, i.e., methanol, toluene, and xylenes, Table 8–1. Accordingly, if possible, avoid storing at any one site an RQ or more of a hazardous substance. If treating chemicals are stored in bulk, the bulk tank maximum level should be established *below* the RQ. However, donít let the regulations control operations. If more than the RQ of a hazardous substance is required, store the required quantity. However, the containers should be fitted with drip pans so releases can be contained. Also, substitution of a non-hazardous substance for a hazardous substance containing chemical serves as a means of avoiding an RQ release.

---

[14] 40 CFR §302.4

[15] 58 FR 54836

[16] 40 CFR §302.6 (a)

[17] 800-424-8802

Table 8–1 List of hazardous substances and reportable quantities

Hazardous Substance	CASRN	Regulatory Synonym	Statutory RCRA Waste No.	Category	Final RQ Pounds(Kg)
Acrolein	07028	2-Propenal	P003	X	1 (0.454)
Ammonia	7664417			B	100(45.4)
Aroclor 1016	2674112	POLYCHLORINATED BIPHENYLS (PCBs)		X	1(0.454)
Aroclor 1221	1104282	POLYCHLORINATED BIPHENYLS (PCBs)		X	1 (0.454)
Aroclor 1232	1141165	POLYCHLORINATED BIPHENYLS (PCBs)		X	1 (0.454)
Aroclor 1242	53469219	POLYCHLORINATED BIPHENYLS (PCBs)		X	1 (0.454)
Aroclor 1248	2672296	POLYCHLORINATED BIPHENYLS (PCBs)		X	1 (0.454)
Aroclor 1254	1097691	POLYCHLORINATED BIPHENYLS (PCBs)		X	1 (0.454)
Aroclor 1260	1096825	POLYCHLORINATED BIPHENYLS (PCBs)		X	1 (0.454)
Asbestos[i]	332214			X	1 (0.454)
Benzene	71432			A	10 (4.54)
Carbon disulfide	75150		P022	B	100 (45.4)
Chlorine	7782505			A	10 (4.54)
Creosote	8001589		U051	X	1 (0.454)
Cresol(s)	319773	Cresylic acid Phenol, methyl	U052	C	1000 (454)
m-Cresol	08394	m-Cresylic acid			
o-Cresol	95487	o-Cresylic acid			
p-Cresol	06445	p-Cresylic acid			
Ethylene glycol	07211			D	5000 (2270)
Formaldehyde	50000		U122	B	100 (45.4)
Lead[ii]	7439921				#
Methanol	67561	Methyl alcohol	U154	D	000 (2270)
Methyl ethyl ketone (MEK)	78933	2-Butanone	U159	D	5000(2270)

Phosphoric acid	7664382			D	5000 (2270)
POLYCHLORINATED BIPHENYLS	1336363			X	1 (0.454)
Aroclor 1016	12674112	POLYCHLORINATED BIPHENYLS (PCBs)			
Aroclor 1221	11104282	POLYCHLORINATED BIPHENYLS (PCBs)			
Aroclor 1232	11141165	POLYCHLORINATED BIPHENYLS (PCBs)			
Aroclor 1242	53469219	POLYCHLORINATED BIPHENYLS (PCBs)			
Aroclor 1248	12672296	POLYCHLORINATED BIPHENYLS (PCBs)			
Aroclor 1254	11097691	POLYCHLORINATED BIPHENYLS (PCBs)			
Aroclor 1260	11096825	POLYCHLORINATED BIPHENYLS (PCBs)			
Pyridine	110861		U196	C	1000 (454)
Sodium chromcte	7775113			A	10 (4.54)
Sodium hydroxide	1310732			C	1000 (454)
Sodium hypochlorite	7681529			B	100 (45.4)
	10022705				
Sulfuric acid	7664939			C	1000(454)
	8014957				
Tetrahydrofuran	109999	Furan, tetrahydro-	U213	C	1000 (454)
Toluene	108883	Benzene, methyl-	U220	C	1000 (454)

i —No reporting of releases of this hazardous substance is required if the diameter of the pieces of the solid metal released is equal to or exceeds 100 micrometers (0.004 inches).

ii — The RQ for asbestos is limited to friable forms only.

# indicates that the RQ is subject to change when the assessment of potential carcinogenicity is completed.

9 Hazard Communication Standard

SUMMARY

The Hazard Communication Standard [HAZCOMM] is not an environmental regulation but a regulation[1] developed under the *Occupational Safety and Health Act* [OSHA].[2] We include a discussion of HAZCOMM because compliance with the *Superfund Amendments and Reauthorization Act* [SARA] depends on an understanding of HAZCOMM.

HAZCOMM represents a means of communicating workplace chemical knowledge to employees. The rule identifies hazardous chemicals and requires preparation of a written HAZCOMM program regardless of the quantity of hazardous chemicals maintained. Employees must be trained in the program and retrained each time a new hazard is introduced to the workplace. In principle, the HAZCOMM concept is simple. However, the definitions are complex and obtuse.

INTRODUCTION

OSHA promulgated the HAZCOMM rule[3] for general industry in 1988 and modified the rule February 9, 1994, at 59 FR 6126 effective

---

[1] 29 CFR §1910.1200

[2] 29 USC 651

[3] 29 CFR §1910.1200

March 11, 1994. HAZCOMM regulations mandate compliance by drilling contractors, service contractors, oil and gas operators, and any other company operating within state boundaries including territorial seas.

If an operator experiences a severe accident requiring an OSHA investigation, OSHA will request information regarding the operator's HAZCOMM program. The EPA automatically investigates a facility where an OSHA investigation is required. Thus, the SARA program and records will be reviewed [see Chapter 10].

To make matters worse, if OSHA or the EPA can prove the operator (1) knowingly and/or willingly contravened the mandate of the rule or (2) knew the regulations required compliance but neglected to implement the program, the operator can be subject to criminal sanctions.

HAZCOMM DEFINITIONS

Hazardous Chemical

Any substance presenting a physical or health hazard or *any mixture* of substances where a hazardous chemical represents 1 percent or more of the mixture [0.1 percent, if a carcinogen][4] *and* stored in a container, except:

- hazardous waste as defined by the RCRA,[5] when subject to regulations issued under that act by the EPA
- tobacco or tobacco products
- wood or wood products
- articles, meaning a manufactured item: (1) which is formed to a specific shape or design during manufacture, (2) which has end use function(s) dependent in whole or in part upon its shape or design during end use, and (3) which does not release, or otherwise result in exposure to, a hazardous chemical, under normal conditions of use
- food, drugs, cosmetics, or alcoholic beverages in a retail establishment that are packaged for sale to consumers

---

[4] 29 CFDR §1910.1200 (c)

[5] 42 USC 6901 et seq.

- foods, drugs, or cosmetics intended for personal use by employees while in the workplace
- any consumer product or hazardous substance, as those terms are defined in the *Consumer Product Safety Act*[6] and *Federal Hazardous Substances Act*[7] respectively, when used in the workplace in the same manner as normal consumer use, and which use results in a duration and frequency of exposure which is not greater than exposure experienced by consumers
- any drug, as that term is defined in the *Federal, Food, Drug, and Cosmetic Act*,[8] when it is in solid, final form for direct administration to the patient [i.e., tablets or pills]

Physical Hazard

The following are considered hazards:[9]

- flammable liquid – flash point <100° F
- combustible liquid – 100° F ≤flash point <200° F
- gas under pressure ≥ 25 psig [compressed gas]
- reactive material
- explosive material – a physical hazard consists of all hazardous materials listed by DOT in 49 CFR §172.101.

Health Hazards

Two types of health hazards exist:

- acute health hazard
- chronic health hazard

Acute health hazard

Any chemical causing immediate adverse health effects by ingestion, inhalation, or contact is an acute health hazard.[10] The measure of

---

[6] 15 USC 2051 et seq.
[7] 15 USC 1261 et seq.
[8] 21 USC 301 et seq.
[9] 29 CFR §1910.1200 (c)
[10] 29 CFR §1910.1200, Appendix A

an acute health hazard depends on the toxic or irritant properties of the chemical.

Toxicity

Toxicity is measured by the lethal dose [LD50] or lethal concentration [LC50] which kills 50 percent of the test animals, typically rats. The LD50 is based on test animal body weight. The LC50 is based on the concentration of the chemical in the atmosphere measured in parts per million [mg/m^3].

Toxic effects occur from ingestion, inhalation, or contact. A chemical is an acute health hazard if:

- oral LD50≤500 mg/kg of specimen body weight[11]
- contact LD50 ≤1,000 mg/kg of specimen body weight[12]
- airborne LC50 ≤2,000 mg/m^3 in the breathing atmosphere[13]

Typically, the operator must rely on the chemical manufacturer for acute health data.

Irritant

A chemical that is not corrosive but causes a reversible inflammatory effect on living tissue by chemical action at the site is a contact toxic chemical. A chemical is a skin irritant if it causes inflammation when tested on the intact skin of albino rabbits by the methods of 16 CFR 1500.41 for four hours exposure or by other appropriate techniques. A chemical is an eye irritant if so determined under the procedure listed in 16 CFR 1500.42 or other appropriate techniques.

- skin irritant score ≥5 using test procedure 16 CFR §1500.41
- eye irritant positive score using test procedure 16 CFR §1500.42

Chronic Health Hazard

Any chemical causing adverse health effects as the result of persistent *prolonged* contact is a chronic health hazard.[14] Any material listed

---

[11] 29 CFR §1910.1200, Appendix A (6) (a)
[12] 29 CFR §1910.1200, Appendix A (6) (b)
[13] 29 CFR §1910.1200, Appendix A (6) (c)
[14] 29 CFR §1910.1200, Appendix A

as a suspected carcinogin or chemical having a listed exposure limit, is a chronic health hazard.

- permissible exposure limit [PEL]
- threshold limit value [TLV]
- time-weighted average [TWA]
- short-term exposure limit [STEL]

Refer to Appendices 5, 6, and 7 for lists of hazardous chemicals that are *chronic health hazards*.[15]

Material Safety Data Sheet

A Material Safety Data Sheet [MSDS] is a manufacturer-prepared written document concerning a chemical.[16] The MSDS should include:

- chemical or mixture name
- chemical or mixture common name
- each individual chemical present at or above 1% [0.1%, if a carcinogen] contributing to a known health hazard or physical hazard
- physical and chemical properties [flash point, vapor pressure] of the chemical or mixture
- health hazards including symptoms or medical signs associated with exposure to the chemical or mixture
- any chronic health hazard threshold limits [PEL, STEL etc.],[17] Appendices 5, 6 and 7
- a listing in the National Toxicology Program Annual Report on carcinogens or as a potential carcinogen in the International Agency for Research on Cancer [ARC] Monographs or by OSHA
- pertinent precautions for safe handling or use
- date of MSDS preparation
- name, address, and telephone number of the chemical manufacturer

---

[15] 29 CFR Section 1910.1000, 58 FR 35338

[16] 29 CFR §1910.1200 (g)

[17] Refer to OSHA Z Tables [29 CFR §1910.1000]

Hopefully, the MSDS provides the user with sufficient information to establish if the product is a hazardous chemical, hazardous substance, or extremely hazardous substance.

Hazard Class

- flammable
- combustible
- reactive
- explosive
- corrosive
- irritant
- carcinogen
- biological agent
- toxic

Container

Any bag, box, sack, drum, tank, can, or reaction vessel is considered a container for purposes of HAZCOMM.[18]

NOTE: piping, separators, heater treaters, and other process vessels are excluded by this definition.

HAZCOMM Requirements

HAZCOMM requirements provide a regulatory mechanism to inform employees of hazards associated with chemicals found in the workplace, i.e., *employee right-to-know*. The standard[19] imposes requirements on all employers manufacturing, storing or using hazardous chemicals regardless of the number of employees or the amount of hazardous chemicals stored. The employer must:

- prepare a written HAZCOMM program[20]
- present the written program to employees in a training session(s)[21]

---

[18] 29 CFR §1910.1200 (c)
[19] 29 CFR §1910.1200
[20] 29 CFR §1910.1200 (e)
[21] 29 CFR §1910.1200 (h)

- collect and retain MSDSs for each hazardous chemical present in the workplace[22]
- prepare a list of hazardous chemicals present in the workplace[23]
- label all containers of hazardous chemicals in the workplace[24]

Written HAZCOMM Program

HAZCOMM requires the owner or operator of a facility, where hazardous chemicals are manufactured, stored or used to prepare, implement, and retain a written program which includes:[25]

- list of hazardous chemicals present in the workplace[26]
- methods used by employer to inform employees of the hazards of non-routine tasks[27]
- methods used by employer to provide other employers on-site access to MSDS for hazardous chemicals in the workplace [multi-employer workplaces][28]
- methods used by employer to provide other employers of precautionary measures needed to protect employees in the multi-employer workplace[29]
- methods used by employer to inform other employers of the labeling system in use at the site[30]

The written HAZCOMM program must be available to employees, their designated representatives, and OSHA. However, the program may be retained at a central workplace facility.

Employee HAZCOMM Information and Training

OSHA requires each employee exposed to a hazardous chemical be provided with an effective training program at the time of their

[22] 29 CFR Section 1910.1200 (g)
[23] 29 CFR §1910.1200 (h) (1) (iii)
[24] 29 CFR §1910.1200 (f)
[25] 29 CFR §1910.1200 (e)
[26] 29 CFR §1910.1200 (e) (1) (i)
[27] 29 CFR §1910.1200 (e) (1) (ii)
[28] 29 CFR §1910.1200 (e) (2)
[29] 29 CFR §1910.1200 (e) (2) (ii)
[30] 29 CFR §1910.1200 (e) (2) (iii)

initial assignment and whenever a new physical or health hazard [in the form of a hazardous chemical] is introduced into the work area.[31]

Information Program

Information made available by the employer to exposed employees should include a description of:

- labels or other forms of warning on containers of hazardous chemicals
- the information required by employers
- operations involving hazardous chemicals
- locations and availability of the written hazard communication program
- list of hazardous chemicals in the workplace
- MSDSs for each hazardous chemical
- labeling of hazardous chemicals

Training Program

The training program given to employees should include:

- methods and observations required to detect the presence of a hazardous chemical
- physical and/or health hazards of the hazardous chemicals
- details contained in the written program

Retain MSDS

HAZCOMM requires each owner or operator of a facility where hazardous chemicals are manufactured, stored, or used to *collect, retain, and make available to employees MSDS for all hazardous chemicals* present in the workplace, i.e., MSDS book.[32] Each hazardous chemical MSDS should be given a unique arbitrary number as discussed below.

[31] 29 CFR §1910.1200 (h)
[32] 29 CFR §1910.1200 (g)

List of Hazardous Chemicals

HAZCOMM requires each owner or operator of a facility where hazardous chemicals are manufactured, stored, or used to prepare a list of *all hazardous chemicals or mixtures* containing 1% (0.1% if a carcinogen) of a hazardous chemical present in the workplace.[33]

Preparing a list of hazardous chemicals presents several concerns that must be considered. Most operators vary hazardous chemical usage with time. For example, the chemicals present at a producing site are different during oil producing operations than during injection operations. Also, different brands of materials can be used at the same fields. Accordingly, the easiest method of preparing the list of hazardous chemicals is to prepare the list alphabetically, Table 9–1.

Table 9–1 Chemical inventory list

CHEMICAL NAME	TYPE OF HAZARD	MSDS NO.
AQUANOX 7EB124	FIRE	1
CARBON DIOXIDE	PRESSURIZED GAS	2
CARBON DIOXIDE	PRESSURIZED GAS	2A
CARBON DIOXIDE	CHRONIC HEALTH	2B
COREXIT 4454	FIRE	3
CRONOX C 6662	ACUTE HEALTH	43
CRONOX C 6668	ACUTE HEALTH	44
CRUDE OIL	FIRE	4
FLUDEX WF0252	CHRONIC HEALTH	5
HYDROCHORIC ACID	ACUTE HEALTH	6
HYDROFLOURIC ACID	ACUTE HEALTH	7
MAGNA-D-WAX 957	FIRE	8
METHANOL	FIRE	9
NALCO VI-SEP 103	FIRE	10
NALCO VI-SEP 4493	FIRE	11
PETRO-SOL 100-C	FIRE	12
PETRO-SOL 500-P	FIRE	13
PRO-TECH E-400	FIRE	14
PRO-TECH E-404	FIRE	15
TECH HIB CS-368	FIRE	16
TECH SOL 500	FIRE	17
TECH TREAT 384	FIRE	18

[33] 29 CFR §1910.1200 (e) (2) (i)

TRETOLITE CLW0016K	FIRE	22
TRETOLITE CRW0010A	FIRE	45
TRETOLITE CRW5400E	FIRE	23
TRETOLITE DMO0077G	FIRE	24
TRETOLITE DMO2241G	FIRE	25
TRETOLITE DMO4024E	FIRE	26
TRETOLITE DMO5122L	FIRE	27
TRETOLITE PAO0023L	FIRE	19
TRETOLITE PAO0048K	FIRE	28
TRETOLITE PAO0072E	FIRE	20
TRETOLITE PAO0100F	FIRE	29
TRETOLITE PAO0103F	FIRE	30
TRETOLITE PAW0039I	CHRONIC HEALTH	31
TRETOLITE RP2327	FIRE	32
TRETOLITE SCW0219K	FIRE	33
TRETOLITE SRW0029C	ACUTE HEALTH	34
TRETOLTE WAW0015F	FIRE	35
VISCO 1111	FIRE	36
VISCO 1153	ACUTE HEALTH	37
VISCO 4493	FIRE	38
VISCO 4772	FIRE	39
VISCO 4902	FIRE	40
XEROX 6R229	CHRONIC HEALTH	41
XEROX 6R234	CHRONIC HEALTH	42

Number each hazardous chemical on the list sequentially. If a new chemical is added, give that chemical the next sequential number but place the chemical in the correct alphabetical order in the list, Table 9–1. As more chemicals are added, the numbers no longer are sequential. However, the chemical can be located easily because chemicals are listed alphabetically but placed in the MSDS book numerically.

If use of a chemical is discontinued, prepare a new list dropping the discontinued number. Place the discontinued MSDS in a book containing discontinued MSDS. However, retain the number on the sheet. If the chemical is used again, transfer the MSDS to the active MSDS book and insert the chemical in the correct alphabetical order in the MSDS list.

Label Hazardous Chemicals

HAZCOMM requires each owner or operator where hazardous chemicals are manufactured, stored, or used to label *all containers of hazardous chemicals* leaving the facility or present at the facility unless the same person filling the container also empties the container.[34]

The manufacturer is require to provide on the container:

- name and address of manufacturer
- name of hazardous chemical
- hazard warning (class)

The user is required to provide on his containers, i.e., tanks, portable containers, etc., the name of the chemical exactly as shown in the MSDS list so the employee can go to the list, find the MSDS by the unique MSDS number, and discover the hazard class or other information.

Multiple Employer Workplace

Oil and gas field operations frequently involve employees from several companies operating at the same workplace, i.e., well pulling crews, roustabouts and cementers at a site. As discussed above, *each employer* is obligated to have an active HAZCOMM program. Additionally, the standard requires each employer having hazardous chemicals onsite to inform the other employers of the hazardous chemicals they have or may bring to the workplace and offer to provide a copy of the MSDS.[35]

Typically, this information exchange is accomplished by mailing a letter to the other employers, i.e., the operator sends a form letter to each roustabout gang, cementer, drilling, and well pulling contractor informing the contractors of the operator's HAZCOMM program, hazardous chemicals present, and the availability of MSDSs. In turn, the contractor is obligated to reciprocate.

HAZCOMM Program Maintenance

HAZCOMM regulations require each operator or manufacturer to supply an MSDS to the purchaser of the hazardous chemical product (crude oil, condensate, LPG) at the time of the *first shipment* and each time the MSDS is modified.[36] *This means producers and drilling contrac-*

[34] 29 CFR §1910.1200 (f)
[35] 29 CFR §1910.1200 (e) (2)
[36] 29 CFR §1910.1200 (g) (6)

tors become manufacturers for the purpose of HAZCOMM, i.e., crude oil, condensate, propane, butanes, natural gasoline, reuse drilling fluids.

Additionally, the MSDS book must be kept current. Accordingly, once per year by the anniversary of the program [May 24], the employer must update the MSDS Book. *This does not require new MSDS but it means each MSDS must be the most current available.*

HOW TO RECOGNIZE A HAZARDOUS CHEMICAL

The HAZCOMM program relies on the employer to have sufficient industrial hygiene knowledge to recognize a hazardous chemical. Unfortunately most small employers do not have the knowledge or funds to hire someone with the knowledge. Further, OSHA offers no help in making the determination.[37] Accordingly, the following description of the properties of a hazardous chemical allows the uninitiated to determine which chemicals are hazardous chemicals:

- First, the existence of an MSDS does not mean the material described is a hazardous chemical.
- Second, sufficient information should be contained in the MSDS to allow recognizing a hazardous chemical from a non-hazardous chemical. Unfortunately, not all MSDS are so complete.

In order to make MSDS more user friendly, OSHA at 56 FR 20580 [May 17, 1990] solicited information regarding a standardized MSDS. As of May 1, 1995, OSHA had not developed a standard MSDS form. Accordingly, industry must work with what is available.

A hazardous chemical is any substance that presents a physical or health hazard.[38]

Physical Hazard

Typically, the easiest hazard determination is the physical hazard. Most MSDS provide fire and explosion hazard data, Figures 9–1, 9–2, 9–3 [Section V] and 9–4, [Section IV].

---

[37] 59 FR 6126, Appendix E, Item 4

[38] 29 CFR §1910.1200 (c)

In the case of chemicals described in Figure 9-1, the flash point equals 111°F. Since the flash point is less than 200°F [FP<200° F], the chemical is a hazardous chemical and represents a combustible liquid [100°F. ≤FP <200° F]. Also, the same determination can be made by observing in Section I, Figure 9–1, the DOT hazard class is "combustible liquid." *Any substance classed as a "hazardous material" by DOT is a physical hazard under the OSHA hazardous chemical classification.* Thus, we conclude the blended polyamide is a combustible liquid, i.e., a hazardous chemical.

Health Hazard

Two distinct health hazards make a substance a hazardous chemical:

- acute health hazard
- chronic health hazard

Acute Health Hazard

MSDS Section VI in Figures 9–1, 9–2, and 9–4, and Section III in Figure 9–3 are designed to provide acute health data. Unfortunately, Section VI provides no indication if the chemical is an acute health hazard, because no LD50 or LC50 values are provided. On the other hand, Figure 9–3 at Section III provides ample health information. However, Section VIII in Figures 9–1 and 9–2 includes Hazardous Material Information System [HMIS] and National Fire Prevention Association [NFPA] ratings for health [acute health].

As discussed below, an HMIS rating of <2 or a NFPA rating of <3 indicates the chemical is not an acute health hazard. Accordingly, the substances described in Section VIII of Figures 9–1 and 9–2 are not acute health hazards.

Hazard Rating Systems

Three hazard rating systems are in use: NFPA, HMIS, and a modified NFPA.

Figure 9–1 Material safety data sheets – manufacturer's information and health hazard data

MATERIAL SAFETY DATA SHEETS

Supplier:
Address:

Emergency

I. - MANUFACTURER'S INFORMATION:

Product Name: CARBO-MIX CAS Number: NAV Code: 1217470
Chemical Name: BLENDED POLYAMIDE
Chemical Description:
A HIGH-ACTIVITY NONIONIC POLYAMIDE EMULSIFIER. AN OIL MUD EMULSIFIER FOR HIGH-WATER SYSTEMS AND ENVIRONMENTALLY SENSITIVE AREAS.

DOT Hazard Class: COMBUSTIBLE LIQUID Shipping Id. No. - UN: NA: 1993

DOT Proper Shipping Name:
COMBUSTIBLE LIQUID, N.O.S., (CONTAINS METHANOL)

II. - INGREDIENTS:

Hazardous Components (OSHA 29 CFR 1910.1200):	PIN:	TLV:	PEL:	%:	CAS Number
METHANOL	NAP	200 PPM	200 PPM	6-8	67-56-1
NAP	NAP	NAP	NAP	NAP	NAP
NAP	NAP	NAP	NAP	NAP	NAP
NAP	NAP	NAP	NAP	NAP	NAP
NAP	NAP	NAP	NAP	NAP	NAP

Hazardous Substance (EPA 40 CFR 302):	PIN:	TLV:	Product RQ:	%:	CAS Number
METHANOL	NAP	200 PPM	5000 GAL	NAP	67-56-1
NAP	NAP	NAP	NAP	NAP	NAP
NAP	NAP	NAP	NAP	NAP	NAP
NAP	NAP	NAP	NAP	NAP	NAP
NAP	NAP	NAP	NAP	NAP	NAP

III. - PHYSICAL DATA:

Boiling Point(F): NAP Vapor Pressure(mmHg): NAP Specific Gravity: 0.95 PH: NAP
Melting Point(F): NAP Vapor Density(Air=1): NAP Percent Volatile By Volume(%): NAP
Freezing Point(F): NAP Solubility In Water: NAP Evaporation Rate: (ACETATE = 1): <1
Odor Threshold: NAP Appearance And Odor: BLACK VISCOUS LIQUID/
Material Is: MIX-LIQUID Coefficient of Water/Oil Distribution: INSOLUBLE IN WATER

IV. - FIRE & EXPLOSION HAZARD DATA:

Flashpoint(F): 111 CLOSED CUP Auto Ignition Temperature(F): NAP
Flammable Limits - LFL: NAP UFL: NAP Explosion - LEL: NAP UEL: NAP
Rate Of Burning: NAP Sensitivity To Impact: NAP
Extinguish Media - Water: CO2: X Halon: Dry Chemical: Foam: X Fog: X
Hazardous Combustion Products: NONE KNOWN
Fire Fighting Procedures:
WEAR POSITIVE PRESSURE SELF-CONTAINED BREATHING APPARATUS.

Unusal Fire and Explosion Hazards:
NONE KNOWN

V. - REACTIVITY DATA:

Chemical Stable: Yes: X No: If no, Under Which Conditions?
HEAT, SPARKS OR OPEN FLAMES.

Incompatiblity (Materials to Avoid):
NAP

Hazardous Decomposition or Byproducts:
NONE KNOWN

Hazardous Polymerization - May Occur: Will Not Occur: X Conditions to Avoid:
NAP

MATERIAL SAFETY DATA SHEETS

Supplier:
Address:

Emergency

VI. - HEALTH HAZARD DATA: CARBO-MIX

Carcinogen - NTP Program: NAP   Carcinogen - IARC Program: NAP

Symptoms of Exposure:
MAY CAUSE SOME EYE IRRITATION IF NOT HANDLED IN A WELL VENTILATED AREA. PROLONGED CONTACT COULD RESULT IN IRRITATION OF THE SKIN.

Medical Conditions Aggravated By Exposure:
NAP

Emergency and First Aid Procedures:
Eyes: FLUSH WITH CLEAN RUNNING WATER FOR AT LEAST FIFTEEN MINUTES AND CONTACT A PHYSICIAN.

Skin: REMOVE CONTAMINATED CLOTHING AND WASH THE AFFECTED AREA WITH SOAP AND WATER.

Swallow: IF CONSCIOUS, GIVE WATER/MILK TO DRINK. DO NOT INDUCE VOMITING. SEEK MEDICAL ATTENTION.

Breath: REMOVE FROM EXPOSURE TO FRESH AIR AND TREAT SYMPTOMATICALLY. CONTACT A PHYSICIAN.

VII. - ENVIRONMENTAL PROTECTION PROCEDURES:

Spill Response:
EXTINGUISH ALL SOURCES OF IGNITIONL. ABSORB LEAKING LIQUID WITH SAND OR EARTH AND DISPOSE OF AS SOLID WASTE.

Waste Disposal Method:
BURY OR INCINERATE IN APPROVED INDUSTRIAL DISPOSAL SITE IN ACCORDANCE WITH LOCAL, STATE AND FEDERAL REGULATIONS.

Handling and Storage:
NAP

VIII. - SPECIAL PROTECTION INFORMATION:

Respiratory Protection: NOT REQUIRED DURING NORMAL HANDLING.
Ventilation: LOCAL EXHAUST
Clothing: NORMAL WORK CLOTHES   Gloves: PVC GLOVES
Eye: SPLASH GOGGLES.
Footwear: NAV   Other: NAP

Known Hazards Under 29 CFR 1910.1200

	Yes		Yes
Combustible Liquid	X	Skin Hazard	X
Flammable Material		Eye Hazard	X
Pyrophoric Material		Toxic Agent	
Explosive Material		Highly Toxic Agent	
Unstable Material		Sensitizer	
Water Reactive Materal		Carcinogen	
Oxidizer		Reproductive Toxin	
Organic Peroxide		Blood Toxin	
Corrosive Material		Nervous System Toxin	
Compressed Gas		Lung Toxin	
Irritant		Liver Toxin	
Nuisance Dust		Kidney Toxin	

HMIS	NFPA
Health: 0	Health: 0
Flammability: 1	Flammability: 1
Reactivity: 0	Stability: 0
Personal Protection: D	Special:

Physical & Health Hazard - SARA 40 CFR Part 370 (Check all that apply) (NOTE (*) indicates TLV for nuisance dust only)
Fire: X   Sudden Release Of Pressure: Reactivity: Immediate(Acute): X   Delayed(Chronic):

IX. - SPECIAL PRECAUTIONS:

Precautions For Repair & Maintenance Of Contaminated Equipment:
NAV

Other Precautions:
NAV

Figure 9–2 Material safety data sheet – drilling fluids

MATERIAL SAFETY DATA SHEET

MATERIAL SAFETY DATA SHEETS

Supplier:
Address:

Emergency

Emergency

I. - MANUFACTURER'S INFORMATION:

Product Name: CAS Number: 14808-60-7 Code: 1215010
Chemical Name: BARIUM SULFATE, BARITE
Chemical Description:
A HIGH PURITY BARITE USED TO INCREASE THE DENSITY OF ALL TYPES OF DRILLING FLUIDS.

DOT Hazard Class: NAP Shipping Id. No. - UN: NA:

DOT Proper Shipping Name:

II. - INGREDIENTS:

Hazardous Components (OSHA 29 CFR 1910.1200):	PIN:	TLV:	PEL:	%:	CAS Number
SILICA (QUARTZ)	NAP	.1 MG/M3	NAP	<5	14808-60-7
NAP	NAP	NAP	**TOTAL DUST	NAP	NAP
NAP	NAP	NAP	NAP	NAP	NAP
NAP	NAP	NAP	NAP	NAP	NAP
NAP	NAP	NAP	NAP	NAP	NAP

Hazardous Substance (EPA 40 CFR 302):	PIN:	TLV:	Product RQ:	%:	CAS Number
SILICA (QUARTZ)	NAP	.1MG/M3	NAP	>5	14808-60-7
**TOTAL DUST	NAP	NAP	NAP	NAP	NAP
NAP	NAP	NAP	NAP	NAP	NAP
NAP	NAP	NAP	NAP	NAP	NAP
NAP	NAP	NAP	NAP	NAP	NAP

III. - PHYSICAL DATA:

Boiling Point(F): NAP
Melting Point(F): 2876
Freezing Point(F): NAP
Odor Threshold: NAP
Material Is: PURE-POWDER

Vapor Pressure(mmHg): NAP
Vapor Density(Air=1): NAP
Solubility In Water: SLIGHTLY
Appearance And Odor: GRAY TO PINK AND ODORLESS
Coefficient of Water/Oil Distribution: INSOLUBLE IN WATER

Specific Gravity: 4.2
Percent Volatile By Volume(%): NAP
Evaporation Rate: (_____ = 1): NAP
PH: INSOLUBLE

IV. - FIRE & EXPLOSION HAZARD DATA:

Flashpoint(F): NAP
Flammable Limits - LFL: NAP UFL: NAP
Rate Of Burning: NAP

Auto Ignition Temperature(F): NAP
Explosion - LEL: NAP UEL: NAP
Sensitivity To Impact: NAP

Extinguish Media - Water: X CO2: X Halon: Dry Chemical: X Foam: X Fog:
Hazardous Combustion Products: WHEN IN PRESENCE OF EXTREME HEAT COULD PRODUCE CO, CO2 FOR THOSE ORES CONTAINING AMTS. OF CARBONATES
Fire Fighting Procedures:
NAP

Unusual Fire and Explosion Hazards:
NAP

V. - REACTIVITY DATA:

Chemical Stable: Yes: X No: If no, Under Which Conditions?
HEATING WITH ALUMINUM CAN PRODUCE AN EXPLOSION.

Incompatiblity (Materials to Avoid):
HEATING WITH ALUMINUUM CAN PRODUCE AN EXPLOSION.

Hazardous Decomposition or Byproducts:
THERMAL DECOMPOSITION PRODUCTS - CO, CO2, OXIDES OF SULFUR

Hazardous Polymerization - May Occur: Will Not Occur: X Conditions to Avoid:
NAP

Supplier:
Address:

MATERIAL SAFETY DATA SHEETS

Emergency Telepnone.

VI. - HEALTH HAZARD DATA:

Carcinogen - NTP Program: NAP Carcinogen - IARC Program: NAP

Symptoms of Exposure:
PROLONGED INHALATION OF THE POWDER CONTAINING FREE CRYSTALLINE SILICA MAY RESULT IN SILICOSIS, A NON-CANCEROUS LUNG DISEASE OR OTHER RESPIRATORY DISEASE.

Medical Conditions Aggravated By Exposure:
NAP

Emergency and First Aid Procedures:
Eyes: FLUSH WITH LARGE AMOUNTS OF CLEAN RUNNING WATER FOR AT LEAST FIFTEEN MINUTES AND CONTACT A PHYSICIAN IF IRRITATION DEVELOPS/PERSISTS.
Skin: WASH WITH SOAP AND WATER

Swallow: CONSULT A PHYSICIAN

Breath: REMOVE FROM EXPOSURE TO FRESH AIR

VII. - ENVIRONMENTAL PROTECTION PROCEDURES:

Spill Response:
NORMAL CLEAN-UP PROCEDURES FOR DRY PRODUCTS - VACUUM, ETC.

Waste Disposal Method:
DISPOSE IN ACCORDANCE WITH LOCAL, STATE AND FEDERAL REGULATIONS.

Handling and Storage:
AVOID CREATING DUST LADEN ATMOSPHERES.

VIII. - SPECIAL PROTECTION INFORMATION:

Respiratory Protection: WEAR NIOSH/MSHA APPROVED RESPIRATOR IF TLV MAY BE EXCEEDED.
Ventilation: LOCAL EXHAUST
Clothing: LAUNDER BEFORE REUSE Gloves: AS PERSONAL NEEDS DICTATE.
Eye: SAFETY GLASSES/GOGGLES
Footwear: NORMAL Other: AS NEEDED FOR PERSONAL COMFORT

Known Hazards Under 29 CFR 1910.1200	Yes		Yes
Combustible Liquid		Skin Hazard	
Flammable Material		Eye Hazard	
Pyrophoric Material		Toxic Agent	
Explosive Material		Highly Toxic Agent	
Unstable Material		Sensitizer	
Water Reactive Materal		Carcinogen	
Oxidizer		Reproductive Toxin	
Organic Peroxide		Blood Toxin	
Corrosive Material		Nervous System Toxin	
Compressed Gas		Lung Toxin	
Irritant		Liver Toxin	
Nuisance Dust	X	Kidney Toxin	

HMIS	NFPA
Health: 1	Health: 1
Flammability: 0	Flammability: 0
Reactivity: 0	Stability: 0
Personal Protection: E	Special:

Physical & Health Hazard - SARA 40 CFR Part 370 (Check all that apply) (NOTE (*) indicates TLV for nuisance dust only)
Fire: Sudden Release Of Pressure: Reactivity: Immediate(Acute): * Delayed(Chronic): . ?

IX. - SPECIAL PRECAUTIONS:

Precautions For Repair & Maintenance Of Contaminated Equipment:
NORMAL

Other Precautions:
AVOID CREATING DUST LADEN ATMOSPHERES.

X. - APPROVAL: Prepared By: Safety Department Code: 1215010 Date Prepared: 05/01/91 Supercedes Issue Date: 08/03/89

Figure 9–3 Material safety data sheet – reproduction machine toner

Material Safety Data Sheet

Date: 9/22/86
Revision: 7/7/92

Manufacturer:

Health Emergency Tel. No.:
Transportation Emergency:
Safety Information Tel. No.:

Section I - Product Identification

Trade Names/Synonyms: Part No.:

Chemical Name: None

Ingredients	CAS No.
Styrene/butadiene copolymer (75-80%)	9003-55-8
Iron oxide (15-20%)	1309-37-1
Rosin acid (<5%)	8050-09-7
Carbon black (<5%)	1333-86-4
Quaternary ammonium salt (<2%)	3843-16-1

Section II - Emergency and First Aid

Eyes:	Flush with water.
Skin:	Wash with soap and water.
Inhalation:	Remove from exposure.
Ingestion:	Dilute stomach contents with several glasses of water.
Primary Route of Entry:	Inhalation
Symptoms of Overexposure:	Minimal respiratory tract irritation may occur as with exposure to large amounts of any non-toxic dust.
Medical Conditions Generally Aggravated by Exposure:	None when used as described by product literature.
Additional Information:	See Sections V and VII. Further information on file in Poisindex.

Section III - Toxicology and Health Information

This material has been evaluated by

Oral LD_{50}:	>10 g/kg (rats) practically non-toxic.
Dermal LD_{50}:	>2 g/kg (rabbits) practically non-toxic.
Inhalation LC_{50}:	>5 mg/l (rats, 4 hr exposure) practically non-toxic.[1] >20 mg/l (rats,calculated 1 hr exposure) non-poisonous, DOT.[1]
Eye Irritation:	Not an irritant.
Skin Sensitization:	Not a sensitizer.
Skin Irritation:	Not an irritant.
Human Patch:	Non-irritating, non-sensitizing
Mutagenicity:	No mutagenicity detected in Ames, Micronucleus, CHO/HGPRT, and Yeast Mitotic Recombination Assays.
Carcinogens:	None present
Aquatic LC_{50}:	>500 mg/l (fathead minnows) non-toxic.

TLV:	10 mg/m³ (total dust)
PEL:	15 mg/m³ (total dust) 5 mg/m³ (respirable dust)
STEL:	None established
Ceiling:	None established
XEL[2]:	2.5 mg/m³ (total dust) 0.4 mg/m³ (respirable dust)

Additional Information: Some of the information reported above is test data of similar products. In a sponsored chronic inhalation study in rats using a special test toner, there were no lung changes at all in the lowest exposure level (1 mg/m³), the most relevant level to potential human exposures. A very slight degree of fibrosis was noted in 25 % of the animals at the middle exposure level (4 mg/m³) while a slight degree of fibrosis was observed at the highest exposure level (16 mg/m³) in all animals. These findings are attributed to "lung overloading," a generic response to excessive amounts of any dust retained in the lungs for a prolonged interval. The special test toner was ten times more respirable than commercially available toner to comply with EPA testing protocol and would not function properly in Xerox equipment.

[1]Based on testing.

Form 58739 (10/84) STAR N.A. - Not Applicable N.E.-None Established N.D. - Not Determined 600E58260

Section IV - Physical Data

Appearance/Odor:	Black powder / faint odor	**Softening Range:**	85°C to 100°C
Boiling Point:	N.A.	**Melting Point:**	N.A.
Solubility in Water:	Negligible	**Specific Gravity (H_2O = 1):**	~1
Evaporation Rate:	N.A.	**Vapor Pressure (mm Hg):**	N.A.
Vapor Density (Air = 1):	N.A.	**pH =**	N.A.

Volatile N.A.%(Wgt.) N.A.%(Vol.)

Section V - Fire and Explosion Data

Flash Point (Method Used):	N.A.	**Flammable Limits**	**LEL:**	N.A.
Extinguishing Media:	Water, dry chemical, carbon dioxide or foam.		**UEL:**	N.A.

Special Fire Fighting Procedures: Avoid inhalation of smoke. Wear protective clothing and self-contained breathing apparatus.

Fire and Explosion Hazards: Toner is a combustible powder. Like most organic materials in powder form, when dispersed in air, it can form explosive mixtures.

Section VI - Reactivity Data

Stability:	Unstable		**Hazardous Polymerization:**	May Occur	
	Stable	X		Will Not Occur	X

Hazardous Decomposition Products: Products of combustion may be toxic. Avoid breathing smoke.

Incompatibility (Materials to Avoid): None known

Section VII - Special Protection Information

Respiratory Protection: None required when used as intended in Xerox equipment.
Eye Protection: None required when used as intended in Xerox equipment.
Protective Gloves: None required when used as intended in Xerox equipment.
Other: For use other than normal customer - operating procedures (such as in bulk toner processing facilities), goggles and respirators may be required. For more information, contact Xerox.

Section VIII - Special Precautions

Handling and Storage: None

Conditions to Avoid: Avoid prolonged inhalation of excessive dust.

Section IX - Spill, Leak, and Disposal Procedures

For Spills or Leakage: Sweep up or vacuum spilled toner and carefully transfer into a sealable waste container. Sweep slowly to minimize generation of dust during clean-up. If a vacuum is used, the motor must be rated as dust tight. A conductive hose bonded to the machine should be used to reduce static buildup (See Section V). Residue can be removed with soap and cold water. Garments may be washed or dry cleaned, after removal of loose toner.

Waste Disposal Method: When disposed, this material is not a hazardous waste according to Federal Regulation 40 CFR 261. However, State and Local requirements may be more restrictive. Therefore, consultation with the appropriate State and Local waste disposal authorities is advised. Incinerate only in a closed container.

Section X - Transportation Information

DOT Proper Shipping Name: Not Regulated
Hazard Classification: N. A. **ID Number:** N. A.

Figure 9–4 Material safety data sheet

Page 1 **MATERIAL SAFETY DATA SHEET**

PRODUCT NAME (Barite)

Section I

Manufacturer:	Telephone Number for Information 713/789-1422
	Emergency Telephone Number 800/424-9300
	Date Prepared 12/08/89

Section II - Hazardous Ingredients/Identity Information

Hazardous Components: Specfic Chemical Identity; Common Name(s)	OSHA PEL	ACGIH TLV	Other Limits Recommended	% (opt.)
Barium Sulfate - $BaSO_4$	No	No	No	

Section III - Physical/Chemical Characteristics

Boiling Point Not Applicable	Specific Gravity (H20 = 1) 4.2-4.6
Vapor Pressure (mm Hg.) Not Applicable	Melting Point Not Applicable
Vapor Density (AIR=1) Not Applicable	Evaporation Rate (Butyl Acetate = 1) N/A
Solubility in Water Insoluble	Appearance and Odor White, gray, red or brown powder-odorless

Section IV - Fire and Explosion Hazard Data

Flash Point (Method Used) Not Applicable	Flammable Limits None	LEL	UEL

Extinguishing Media Not Applicable

Special Fire Fighting Procedures None

Unusual Fire and Explosion Hazard Non-combustible

Section V - Reactivity Data

Stability	Unstable		Conditions to Avoid None
	Stable	X	

Incompatibility (Materials to avoid) Aluminum, Potassium, Magnesium, Heat

Hazardous Decomposition or Byproducts None

Hazardous Polymerization	May Occur		Conditions to Avoid None
	Will Not Occur	X	

Page 2 **MATERIAL SAFETY DATA SHEET**

PRODUCT NAME (Barite)

Section VI - Health Hazard Data

Route(s) of Entry: Inhalation? Yes Skin? Yes Ingestion? Yes

Health Hazards (Acute and Chronic) None

Carcinogenicity: NTP? No IARC Monographs? No OSHA Regulated? No

Signs and Symptoms of Exposure Sneezing, coughing, watery eyes

Medical Conditions Generally Aggravated by Exposure Asthma, Bronchitis from dust

Emergency and First Aid Procedures

Eyes: Flush with water immediately

Skin: Wash with water immediately

Inhalation: Remove to fresh air if ill effects occur. If irritation develops, consult a physician

Ingestion: If large amounts are swallowed, induce vomiting and seek medical attention

Section VII - Precautions for Safe Handling and Use

Steps to Be Taken in Case Material is Released or Spilled Vacuum or sweep up and return to container for reuse or discard

Waste Disposal Method Sanitary landfill

Precautions to Be Taken in Handling and Storing Not classified as a hazardous material. Store in dry area.

Other Precautions Keep dust to a minimum

Section VIII - Control Measures

Respiratory Protection (Specify Type) Wear NIOSH approved dust mask

Ventilation	Local Exhaust Recommended	Special None
	Mechanical (General) Recommended	Other None

Protective Gloves Not Necessary | Eye Protection Safety Glasses

Other Protective Clothing or Equipment None Required

Work/Hygienic Practices Wash with soap and water

National Fire Prevention Association Rating System

The NFPA rating system includes three hazard classes:

- reactivity
- fire
- health [acute health]

Additionally, the system includes special provisions, i.e., use no water [W] etc., Figure 9–5.

Figure 9–5 NFPA rating system

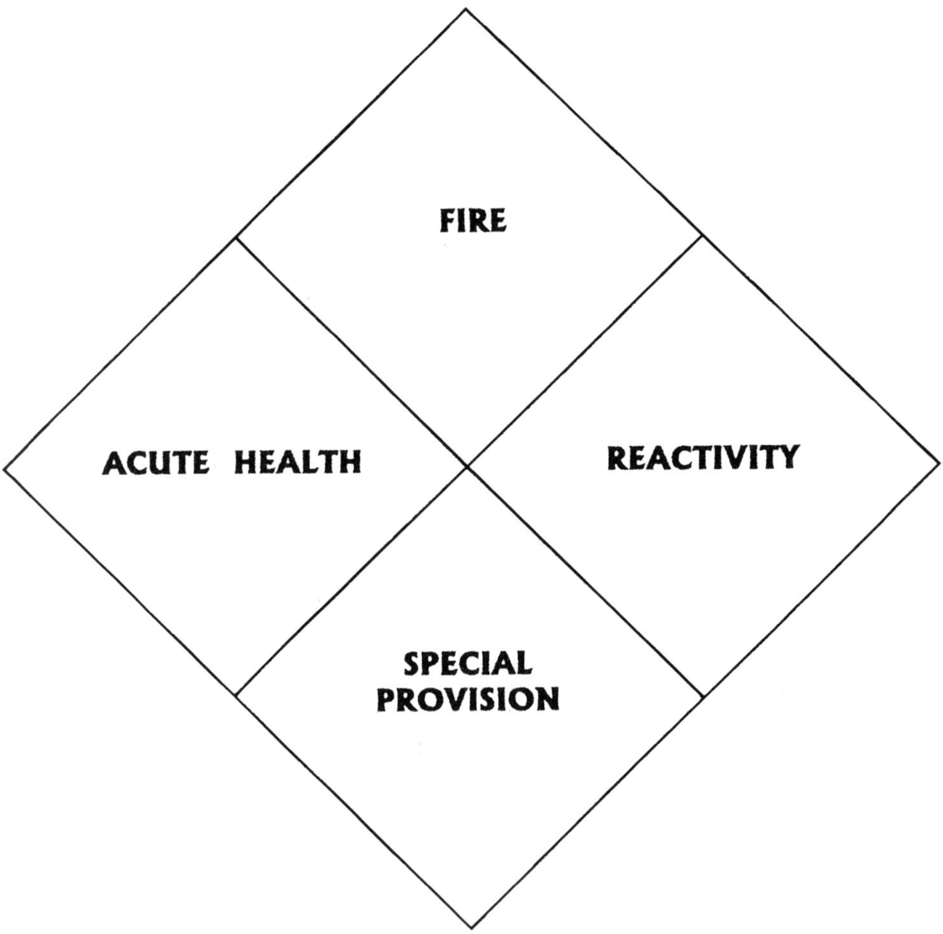

NOTE: The NFPA rating system is silent regarding "chronic health" effects. The NFPA system rates each hazard from 0-4. Under this system a substance is a hazardous chemical if:

reactivity = 1 or more [R ≥1]
fire = 2 or more [F ≥2]
health [acute health] = 3 or more [H≥3]

Hazardous Materials Information System

The HMIS rating system includes the same three hazard classes as the NFPA:

- reactivity
- fire
- health [acute health]

Additionally, the system includes personal protection, Figure 9–1 Section VIII. The rating system includes "chronic health" by placing an asterisk (*) after the acute health rating but was not used by the preparer of Figures 9–1 or 9–2. The HMIS system rates each hazard from 0-4. Under this system a substance is a hazardous chemical if:

reactivity = 1 or more [R ≥1]
fire = 2 or more [F ≥2]
health [acute health][39] = 2 or more [H ≥2].

Modified NFPA Rating System

Several manufacturers use a modified NFPA rating that avoids use of the diamond. For example, the MSDS for barite, Figure 9–6 includes a modified NFPA system along the edge of the MSDS. However, the same NFPA rating system applies. However, the information is presented as a series of boxes containing numbers ranging from 0-4. Based on the modified rating system, barite, Figure 9–6, is not a reactive [R <1], fire [F <2], or acute health [H <3] hazard. However, barite is a chronic health hazard because of silica, Figure 9–6, Section V, and therefore a hazardous chemical. This demonstrates a shortcoming of the short-cut rating systems, i.e., *no chronic health rating.*

---

[39] Add an asterisk (*) after the acute health rating to signify a chronic health hazard.

Figure 9–6 Environmental, safety, and transportation data sheet

I PRODUCT IDENTIFICATION	
SUPPLIER	REGULAR TELEPHONE NO. EMERGENCY TELEPHONE NO.
ADDRESS	
TRADE NAME	
GENERIC DESCRIPTION BARITE, BARIUM SULPHATE, BARYTES	

II HAZARDOUS INGREDIENTS		
MATERIAL OR COMPONENT		HAZARD DATA
SILICA 7631-86-9	2-6	LOW CONCENTRATIONS OF
		CRYSTALLINE SILICA (SiO_2)
		IN THE FORM OF QUARTZ,
		CRISTOBALITE, AND TRIDYMITE
		MAY BE PRESENT
		(SEE SECTION V)

III PHYSICAL DATA		
BOILING POINT (°F) NA	MELTING POINT 2901°F	FREEZING POINT NA
SPECIFIC GRAVITY (H_2O = 1) 4.2	VAPOR PRESSURE (mm Hg) NA	
VAPOR DENSITY (AIR = 1) NA	SOLUBILITY IN H_2O, % BY WT. LOW	
% VOLATILES BY VOL. NA	EVAPORATION RATE (BUTYL ACETATE = 1) NA	
APPEARANCE AND ODOR WHITE TO GRAY TO TAN POWDER, ODORLESS	Density @ 20°C: 4.5@15°C	
pH ND		

HEALTH HAZARD	FLAMMABILITY	REACTIVITY
2	0	0

Ratings based on NIOSH "Identification System for Occupationally Hazardous Materials" (1974)

N/A = Not Applicable N/D = Not Determined

All information recommendations and suggestions herein concerning our product are based upon tests and data believed to be reliable, however, it is the user's responsibility to determine the safety, toxicity, and suitability for his own use of the product described herein. Since the actual use by others is beyond our control no guarantee, expressed or implied, is made by ... Inc. as to the effects of such use, the results to be obtained, or the safety and toxicity of the product nor does NL Petroleum Services, Inc. assume any liability arising out of use, by others, of the product referred to herein. Nor is the information herein to be construed as absolutely complete since additional information may be necessary or desirable when particular or exceptional conditions or circumstances exist or because of applicable laws or government regulations.

IV FIRE AND EXPLOSION DATA

THIS PRODUCT IS NOT FLAMMABLE OR EXPLOSIVE.

EXTINGUISHING MEDIA: WATER, FOAM, CARBON DIOXIDE

V HEALTH HAZARD INFORMATION

CARCINOGENICITY — SEE ROUTES OF EXPOSURE AND EFFECTS (BELOW)

ACUTE ORAL LD_{50}	ACUTE DERMAL LD_{50}	AQUATIC TOXICITY (LC_{50})
ND	ND	7500 ppı

ROUTES OF EXPOSURE AND EFFECTS

THIS PRODUCT CONTAINS FREE CRYSTALLINE SILICA WHICH ACCORDING TO THE IARC HAS EXHIBITED LIMITED EVIDENCE OF CARCINOGENICITY IN HUMANS. PROLONGED INHALATION OF THE POWDER MAY RESULT IN SILICOSIS, A NONCANCEROUS LUNG DISEASE.

TLV FOR RESPIRABLE DUST

$$\frac{10\ MG/M^3}{\%\ RESPIRABLE\ QUARTZ + 2}$$

TLV FOR "TOTAL DUST"

$$\frac{30\ MG/M^3}{\%\ QUARTZ + 3}$$

IF CRISTOBALITE OR TRIDYMITE IS DETECTED, USE ONE-HALF THE VALUE CALCULATED FROM FORMULAE FOR QUARTZ

SKIN: POTENTIAL IRRITANT; PROLONGED CONTACT MAY CAUSE DERMATITIS
EYES: IRRITANT
INHALATION: IRRITATION TO LUNGS, NOSE, AND THROAT; PROLONGED INHALATION MAY CAUSE LUNG INJURY OR DISEASE

EMERGENCY AND FIRST AID PROCEDURES

EYE: IRRIGATE IMMEDIATELY WITH WATER. IF IRRIGATION PERSISTS, CONTACT PHYSICIAN.
FLUSH SKIN WITH WATER.

SWALLOW: GIVE WATER AND INDUCE VOMITING.

IF INHALED IN LARGE AMOUNTS MOVE TO FRESH AIR.

VI REACTIVITY DATA
CONDITIONS CONTRIBUTING TO INSTABILITY STABLE
INCOMPATIBILITY NONE
HAZARDOUS DECOMPOSITION PRODUCTS NONE
CONDITIONS CONTRIBUTING TO HAZARDOUS POLYMERIZATION WILL NOT OCCUR.
VII SPILL OR LEAK PROCEDURES
STEPS TO BE TAKEN IF MATERIAL IS RELEASED OR SPILLED NORMAL HOUSEKEEPING, SWEEP UP AND SALVAGE.
NEUTRALIZING CHEMICALS NA
WASTE DISPOSAL METHOD DISPOSE OF IN ACCORDANCE WITH LOCAL, STATE, AND FEDERAL REGULATIONS.
VIII INDUSTRIAL HYGIENE CONTROL MEASURES
VENTILATION REQUIREMENTS MECHANICAL, GENERAL ROOM VENTILATION USE LOCAL VENTILATION TO MAINTAIN TLV (SEE SECTION V)
SPECIFIC PERSONAL PROTECTIVE EQUIPMENT
RESPIRATORY USE A NIOSH APPROVED MECHANICAL FILTER RESPIRATOR FOR NONTOXIC DUSTS, IF DUST CONCENTRATION EXCEEDS 10mg/m^3
EYE GOGGLES
GLOVES WORKING GLOVES
OTHER CLOTHING AND EQUIPMENT APRON, EYEWASH STATION

IX SPECIAL PRECAUTIONS

PRECAUTIONARY STATEMENTS

DO NOT INGEST
AVOID INHALATION AND PROLONGED SKIN CONTACT

RECOMMENDED LABELING:

FRONT PANEL: CAUTION
SEE BACK PANEL FOR CAUTION BEFORE USE.

BACK PANEL: CAUTION
THIS PRODUCT CONTAINS FREE CRYSTALLINE SILICA WHICH ACCORDING TO THE IARC HAS EXHIBITED LIMITED EVIDENCE OF CARCINOGENICITY IN HUMANS. PROLONGED INHALATION OF THE POWDER MAY RESULT IN SILICOSIS, A NONCANCEROUS LUNG DISEASE. AVOID CREATING DUSTY CONDITIONS AND USE A NIOSH APPROVED DUST RESPIRATOR.

OTHER HANDLING AND STORAGE REQUIREMENTS

STORE IN SHELTERED AREA OR COVER FOR MOISTURE PROTECTION

DEPARTMENT OF TRANSPORTATION INFORMATION

PROPER SHIPPING NAME: NOT REGULATED

PLACARDS: NONE

REPORTABLE QUANTITY: NONE

HAZARD CLASS: NOT HAZARDOUS

ID NUMBER: NONE

HAZARDOUS SUBSTANCE: NONE

LABEL: NONE REQUIRED

CHRONIC HEALTH HAZARD

Any mixture containing 1% or more by weight of a chemical having any exposure limit, i.e. PEL, TVL, TWA, STEL, is a chronic health hazard and thus a hazardous chemical. Section II in Figures 9–1, 9–2; Section III in Figure 9–3, and Section V in Figure 9–6 provide TLV and PEL values for components that constitute >1% by weight of the product. Accordingly, all of the substance described in Figures 9–1, 9–2, 9–3, and 9–6 are chronic health hazards and therefore hazardous chemicals.

Looking specifically at barium sulfate, Figure 9–2, will find a TLV of 0.1 mg/M^3 for silica [Section II] that is less than 5% of the product. Accordingly, barium sulfate is a hazardous chemical.

Similarly, reproduction machine toner, Figure 9–3, contains hazardous chemicals [iron oxide and carbon black] in excess of 1%, both chemicals have PEL and TWA values [Section III]. Accordingly, toner is a hazardous chemical requiring a HAZCOMM program.

MSDS Problems

MSDS, in many cases, lack sufficient data, present erroneous data or conflicting data.

Lack of Data

For example, Figure 9–4, an MSDS for barium sulfate states, at Section II, barium sulfate has no PEL, TLV or "other limits recommended." On the other hand, Figure 9–2 shows barium sulfate contains about 5% silica [Section II], having a TLV of 0.1 mg/m^3. Another manufacturer, Figure 9–6, Section V, provides the same information but in slightly different form. The lack of requisite data for the Figure 9–4 product means a proper HAZCOMM determination cannot be made by the operator from the MSDS alone.

Because the operator is not a trained hygienist and cannot be expected to test the product for hazardousness, I recommend the operator rely implicitly on the manufacturers' MSDS. In this case, barium sulfate purchased from the Figure 9-4 supplier is not a hazardous chemical. However, identical barium sulfate purchased from Figure 9–2 and 9–6 suppliers would be a hazardous chemical.

Erroneous Data

Section VIII, Figures 9–1 and 9–2, provides HMIS and NFPA data. The health rating is <2. This is correct for acute health, however, blended polyamide and barium sulfate are chronic health hazards. Accordingly, an asterisk (*) should be placed next to the health rating of the HMIS system.

Further, in Figure 9–1, Section VIII, the flammability rating for HMIS and NFPA equals 1, meaning the flash point of blended polyamide is ≥200° F. However, Section IV, Figure 9–1 gives the flash point as 111° F. Accordingly, the HMIS and NFPA ratings should be 2, i.e., <200°F.

PROBLEM AREAS OF HAZCOMM

Typically, the most difficult part and most often out of compliance area of hazcomm is container labeling. There is *no one way* to accomplish the labeling requirement.

The concept is simple, i.e., be able to go from the label on the container to the product name and product hazard class. Some operators place the NFPA diamond, Figure 9–5, on the container. In my opinion, the NFPA diamond is not adequate because *no chronic health hazard rating is included* in the NFPA rating system. Other operators place all of the information on the container using stencils. Others number the container and place the same number on the MSDS. For example, all crude oil tanks can be numbered 1. Also, the crude oil MSDS is numbered 1. This is a convoluted labeling system.

The most popular labeling system relies on labeling the container with the name of the product. With this information, the employee can go to the hazardous chemical list, Table 9–1, to determine the hazard class and the MSDS number. A more complete description of the product and first aid data are available from the referenced MSDS.

10 Superfund Amendments and Reauthorization Act

SUMMARY

The *Superfund Amendments and Reauthorization Act* [SARA] provides the community with information regarding the presence of hazardous chemicals at a facility within the state or county. Three classes of chemicals are included:

- extremely hazardous substances
- hazardous substances [Chapter 8]
- hazardous chemicals [Chapter 9]

SARA requires notification and inventorying if the threshold planning quantity [TPQ] or 500 lb of an extremely hazardous substance or 10,000 lb of a hazardous chemical are present at a facility. Additionally, a facility emergency coordinator must be appointed, if extremely hazardous substances at or above the TPQ are present at the facility. The coordinator is responsible for reporting releases of extremely hazardous and hazardous substances.

E&P operators must submit an annual inventory of hazardous chemicals on or before March 1 of each year.

INTRODUCTION

SARA was enacted as an extension of CERCLA. However, its official name is *Emergency Planning and Community Right-to-Know*

Act[1] [EPCRA]. Several sections of the act impose requirements on E&P operators. These include:

- Section 302 – extremely hazardous substance notification
- Section 303 – facility emergency coordinator appointment
- Section 304 – extremely hazardous and hazardous substance release reporting
- Section 311 – hazardous chemical notification
- Section 312 – hazardous chemical inventory
- Section 327 – transportation exemption

None of the SARA provision are difficult. However, they are time consuming.

SARA DEFINITIONS

- extremely hazardous substance – a substance listed in 40 CFR Part 355 Appendix A [Appendix 8]
- hazardous substance – a substance listed in 40 CFR §302.4, Table 302.4 [Appendix 4]
- facility – all buildings, equipment, structures, and other stationary items located on a single site or on contiguous or adjacent sites owned or operated by a single company or operator
- release – spilling, leaking, pumping, pouring, emitting, emptying, discharging, injecting, etc. of any hazardous chemical, hazardous substance, or extremely hazardous substance
- state emergency response commission [SERC]– a state agency designated by the Governor – typically, the state health department
- local emergency planning committee [LEPC] – a local county committee formed by the state- appointed chairman for industry members subject to Section 303
- fire Department – the local fire department or fire departments serving a facility during an emergency
- threshold planning quantity [TPQ] – a quantity of an extremely hazardous substance present at a facility at and above which notification under 40 CFR Part 355 is required [TPQs are listed in 40 CFR Part 355, Appendix A (Appendix 8).]
- reportable quantity [RQ] – a quantity of an extremely hazard-

[1] 42 USC 11001 et seq.

ous substance or hazardous substance at and above which, if released to air, soil, or water, must be reported. [RQs are listed in 40 CFR Part 302 and 355 for hazardous substances and extremely hazardous substances, respectively (Table 8–1, Appendix 8).]

- emergency coordinator – an industry representative of a facility having at or above the TPQ of an extremely hazardous substance at a facility
- Tier Two Cover Sheet – a Texas required form transmitting Section 312 inventory data to agencies, committees, and fire departments
- Tier Two Form – a federal form used for submitting annual inventory data under Section 312 [Figure 10-1] [In Louisiana only the state tier two form can be used.]

SARA Requirements

SARA constitutes the *community right-to-know* program resulting from the 1984 Bhopal, India disaster when more than 2,000 people were killed in a chemical gas release from a Union Carbide insecticide plant. Six sections of the statute impact oil and gas operators: These are:

- Section 302 – notification of the presence of an extremely hazardous substance at or above the TPQ,[2] Appendix 8.
- Section 303 – appointment of an emergency coordinator[3]
- Section 304 – notification in the event of a release of a hazardous substance or extremely hazardous substance at or above the RQ[4]
- Section 311 – submission of a list or individual MSDS for all hazardous chemicals present at a facility at or above 10,000 pounds or extremely hazardous substance at or above TPQ or 500 pounds whichever is less[5]

---

2 42 USC 11002 and 40 CFR §355.30 (b)

3 42 USC 11003 and 40 CFR §355.30 (c)

4 42 USC 11004 and 40 CFR §355.40

5 42 USC 11021 and 40 CFR §370.20

Figure 10–1 Tier two form

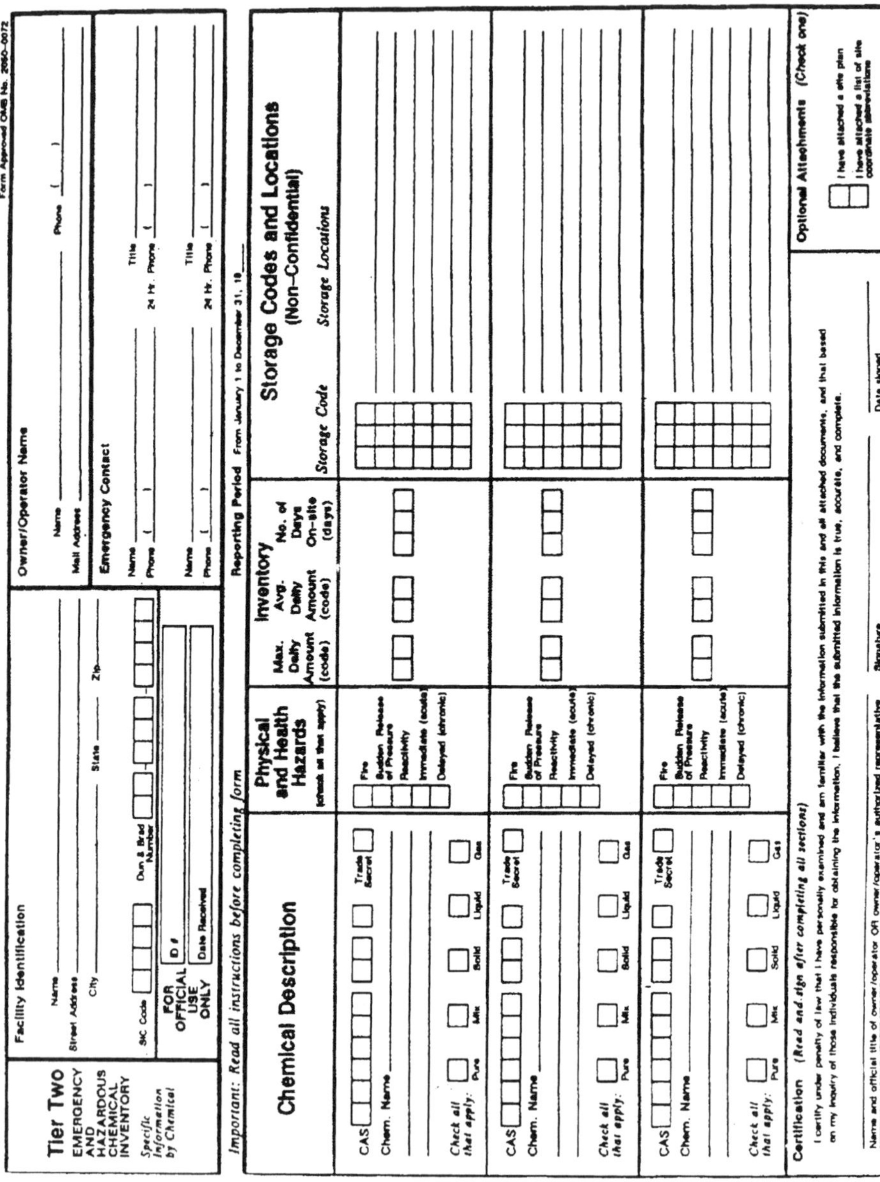

Page ____ of ____ pages
Form Approved OMB No. 2050–0072

Tier Two
EMERGENCY AND HAZARDOUS CHEMICAL INVENTORY
Specific Information by Chemical

Facility Identification
Name ____
Street Address ____
City ____ State ____ Zip ____
SIC Code ____ Dun & Brad Number ____

FOR OFFICIAL USE ONLY: ID # ____ Date Received ____

Owner/Operator Name
Name ____ Phone ( )
Mail Address ____

Emergency Contact
Name ____ Title ____
Phone ( ) 24 Hr. Phone ( )
Name ____ Title ____
Phone ( ) 24 Hr. Phone ( )

Important: Read all instructions before completing form

Reporting Period From January 1 to December 31, 19____

Chemical Description	Physical and Health Hazards (check all that apply)	Inventory: Max. Daily Amount (code)	Inventory: Avg. Daily Amount (code)	Inventory: No. of Days On-site (days)	Storage Codes and Locations (Non-Confidential): *Storage Code*	*Storage Locations*
CAS ____ Trade Secret ☐ Chem. Name ____ *Check all that apply:* ☐ Pure ☐ Mix ☐ Solid ☐ Liquid ☐ Gas	☐ Fire ☐ Sudden Release of Pressure ☐ Reactivity ☐ Immediate (acute) ☐ Delayed (chronic)					
CAS ____ Trade Secret ☐ Chem. Name ____ *Check all that apply:* ☐ Pure ☐ Mix ☐ Solid ☐ Liquid ☐ Gas	☐ Fire ☐ Sudden Release of Pressure ☐ Reactivity ☐ Immediate (acute) ☐ Delayed (chronic)					
CAS ____ Trade Secret ☐ Chem. Name ____ *Check all that apply:* ☐ Pure ☐ Mix ☐ Solid ☐ Liquid ☐ Gas	☐ Fire ☐ Sudden Release of Pressure ☐ Reactivity ☐ Immediate (acute) ☐ Delayed (chronic)					

Certification *(Read and sign after completing all sections)*
I certify under penalty of law that I have personally examined and am familiar with the information submitted in this and all attached documents, and that based on my inquiry of those individuals responsible for obtaining the information, I believe that the submitted information is true, accurate, and complete.

Name and official title of owner/operator OR owner/operator's authorized representative ____ Signature ____ Date signed ____

Optional Attachments *(Check one)*
☐ I have attached a site plan
☐ I have attached a list of site coordinate abbreviations

- Section 312 – submission of an annual inventory of hazardous chemicals or extremely hazardous substances reported under Section 311[6]
- Section 327 – exemption from Section 311 notification and Section 312 inventory requirements when the hazardous chemicals are covered by active DOT shipping papers.

Extremely Hazardous Substance Notification

SARA requires the owner or operator to notify the state emergency response commission [SERC] on or before May 17, 1987, or within 60 days after a facility becomes subject to emergency planning notification, i.e., if the owner or operator stores or has present at the facility a quantity of a liquid extremely hazardous substance [EHS] at or above the TPQ, Appendix 8.[7]

The TPQ is calculated on the quantity of pure extremely hazardous substance present. First, for mixtures, the concentration of extremely hazardous substance must be 1% or more [0.1% for a carcinogen] or the mixture is exempt. Second, a mixture containing an extremely hazardous substance would be reported if the mass of the pure extremely hazardous substance exceeds the TPQ, i.e., the total mass of the mixture need not be reported.[8]

For solids, two criteria apply. These criteria are differentiated from liquids in Appendix 8 by means of two TPQ values separated by a forward slash, i.e., acetone thio-semicarbonized 1,000/10,000. The lower TPQ [i.e., 1,000 lb] applies if the solid exists in powdered form having a particle size less than 100 micron or is handled in solution or in molten form or has an NFPA reactivity rating of 2 or more, i.e., a reactive solid [Chapter 9]. If the solid does not meet any of these criteria, the TPQ for solid equals 10,000 pounds.

Typically, E&P operators are not subject to emergency planning notification because few EHS are present at oil and gas facilities. However, some gas plants become subject to Section 302 notification because of the presence of:

- Acrolein TPQ = 500 lb
- Formaldehyde TPQ = 500 lb
- Sulfuric acid TPQ = 1,000 lb

---

[6] 42 USC 11022 and 40 CFR §370.25

[7] 40 CFR §355.30 (b) and 42 USC 11002

[8] 40 CFR §355.30 (e)

Hydrogen sulfide in the reservoir is exempted from notification. The notification can simply be a letter to the state emergency response Commission stating the facility located at a particular site stores the extremely hazardous substance(s) at or above the threshold planning quantity. Only one notification is required per chemical.

However, the operator must update the information if a new EHS is stored onsite.[9] Additionally, upon request of the local emergency planning committee [LEPC], the operator must provide information relating to emergency planning for the EHS.[10]

Emergency Coordinator

Any owner or operator of a facility subject to Section 302, above, must appoint, within 30 days of becoming subject to these regulations, a facility emergency coordinator who will participate with other emergency coordinators in LEPC activities, i.e., evacuation or other contingency plans in the event of a release of EHS.

Additionally, the emergency coordinator or his alternate must report releases of EHS, Appendix 8, and hazardous substances at or above the RQ.[11]

Release Notification

Extremely Hazardous Substance

A release *that leaves the boundaries of the facility at or above the RQ* must be reported immediately to the LEPC and the SERC.[12] The notification must include:[13]

- chemical name
- a determination if the chemical is an EHS
- estimated quantity released
- time and duration of release
- media or medium into which chemical was released

---

9 40 CFR §355.30 (d) (1)
10 40 CFR §355.30 (d) (2)
11 40 CFR §355.30 (c)
12 40 CFR §355.40
13 40 CFR §355.40 (b)

- known health impact
- precautions to be taken, i.e., evacuation
- name and telephone number of person making the report

As soon as possible, provide a written confirmation.

Hazardous Substance

Within the boundaries of the facility, report the release of a hazardous substance at or above the RQ [Appendix 4] to the National Response Center [1-800-424-8802] under provisions of CERCLA. Outside the boundaries of the facility, report as required for an EHS.

Exemptions

The release reporting does not apply to:

- a release if the EHS is not a hazardous substance and affects only persons within the boundaries of the facility, i.e., within the lease boundaries; however, because most EHS are hazardous substances, CERCLA reporting is required [Chapter 8]
- any release federally permitted:
  - — NPDES permit [Chapter 2]
  - — hazardous waste treatment, storage, or disposal permit [Chapter 7]
  - — ocean dumping[14]
  - — underground injection control [Chapter 3]
  - — air permits [Chapter 1]
  - — *Atomic Energy Act*
- any continuous and stable release[15]

HAZARDOUS CHEMICAL REPORTING

All owners or operators of a facility required to comply with OSHA HAZCOMM [Chapter 9] must provide MSDS or a list of hazardous chemicals to the SERC, LEPC and the local fire department [serving the facility] if at *any one time 10,000 lb or more* of a hazardous

---

[14] 33 USC 1412a

[15] 40 CFR §302.8

chemical or *500 lb or more or TPQ or more* [whichever is least] of an EHS is present at the facility. Notification must be made within 90 days of becoming subject to Section 311.[16] This provision currently applies in all states.

As with HAZCOMM, a mixture containing 1% or more of a hazardous chemical is a hazardous chemical [mixtures containing less than 1% of a hazardous chemical are exempt from HAZCOMM and SARA]. The mass of the mixture or the pure component may be considered when determining if 10,000 lb or more of a hazardous chemical or 500 lb or TPQ of EHS are present. Typically, operators are better served by calculating the amount of hazardous chemical present on the basis of pure component, i.e., hydrochloric acid in a 15% frac fluid.[17] However, if the hazardous chemical is an EHS, the operator must aggregate the quantities of each separate EHS in all mixtures to determine if the 500 lb or TPQ is equaled or exceeded.[18]

As a practical matter, the LEPC, SERC, and fire department object to receiving MSDS in lieu of a list of hazardous chemicals when receiving a Section 311 notification. Accordingly, I recommend submitting a letter of notification and a list of the chemicals. Provide the principle hazard class for each chemical included with the notification.

Remember, the 10,000 lb, 500 lb or TPQ must be present at any one time – do not accumulate over a period of time the hazardous chemicals present at the facility. Do not add different hazardous chemicals together to make the 10,000 lb or the TPQ.

The secret to Section 311 compliance is carefully crafting the geography of the facility. Typically, placing all leases within a field into a single facility reduces reporting requirements. Also, monitoring inventories so no more than 9,990 lb of a single hazardous chemical is present at a facility at any time avoids inclusion of that chemical under Section 311. It is obviously impossible to reduce crude oil to less than 10,000 lb [20 bbl]. Accordingly, *every tank battery must be included in the Section 311 notification.* Also, most drilling rigs store more than 1,300 gallons (10,000 lb) of diesel fuel. Thus, drilling rigs typically are subject to SARA Section 311 notification. *The operator of the lease is responsible for diesel fluid and drilling fluid component notification under Section 311.*

---

[16] 40 CFR §370.20 (b) (1)

[17] 40 CFR §370.28

[18] 40 CFR §370.28 (c)

There is one forgiving feature of Section 311, i.e., once a hazardous chemical notification is made, *no supplemental notification is required for that facility*. However, if a new chemical is used at or above the threshold quantity, notification for the new chemical is required within 90 days. Notification can be made by letter.

EPA Hazard Classes

OSHA provides an extensive list of hazard classes [Chapter 9.] On the other hand, the EPA limits the hazard class under SARA to five classes:[19]

- fire [FP<200° F]
- sudden release of pressure [pressure >25 psig]
- reactivity [violent reaction when mixed with water]
- acutehealth hazard [refer to Chapter 9 for definition]
- chronic health hazard [any chemical having an adverse effect under long term exposure]

NOTE: Chronic health hazard does not include "comfort class" PEL, so lubricating oil is not a SARA hazardous chemical.

INVENTORY OF HAZARDOUS CHEMICALS

Any hazardous chemical notification under Section 311 triggers the requirement to submit an annual inventory on or before March 1 of each year if 10,000 lb of the hazardous chemical or 500 lb or TPQ of an EHS is stored onsite at any time during the previous calendar year.[20] *The Section 312 threshold in Louisiana is 500 lb*. Thus, some Louisiana facilities not subject to Section 311 must submit a Section 312 inventory. Similarly, Louisiana requires use of the Louisiana tier two form for Section 312 submissions.

Texas requires the inventory be submitted on tier two forms, preferably on Texas Continuation forms. Most other states accept the federal tier two form [Figure 10–1] In either case, the information must be submitted with a Texas tier two cover sheet plus an annual filing fee.

---

[19] 40 CFR §370.40
[20] 40 CFR §370.25

Fundamentally, the inventory includes:

- name of hazardous chemical
- CAS number [Table 10–1]
- hazard class of chemical
- form of chemical
- maximum amount present at any time at the facility
- average amount present at any time at the facility
- time chemical present at facility
- storage method
- storage temperature
- storage pressure
- storage location

Table 10–1 Chemical Abstract Service registry number for E&P hazardous chemicals

acrolein	107028
n-butane	106-97-8
propane	74-98-6
ethane	74-84-0
i-butane	75-28-5
raw gasoline mix	C4 C8 8006-61-9
	C5 C8 68425-31-0
residue gas	8006-14-2
crude oil	68410-00-4

The form does not require many specifics. Instead, most responses to the tier two Form are in the form of codes having broad ranges, Table 10–2, 10–3, and 10–4. For example, a 16,000 gal diesel fuel tank contains about 120,000 lb. This fits into quantity range value 05 [Table 10–2; 100,000 to 999,999 lb]. Similarly, most crude oil tanks range from 210 bbl to 1,000 bbl. For a 210 bbl tank, the full contents weight equals about 61,000 lb. or range value 04. The full contents weight of a 1,000 bbl tank equals about 289,000 lb. or range value 05. Accordingly, most Section 312 inventory hazardous chemicals range from 04 to 05.

Other Section 312 inventory entries use codes, i.e., storage type, Table 10–3 and storage conditions, Table 10–4. As with SARA Section 311 notification, the Section 312 inventory for hazardous chemicals can be based on the mass of the mixture or the individual hazardous chemical. However, if you notify under Section 311 on the basis of mixtures, then you must use mixtures when preparing the Section 312 inventory.

Table 10–2 SARA inventory ranges

Range value	Weight range in pounds From	Weight range in pounds To
01	0	99
02	100	999
03	1,000	9,999
04	10,000	99,999
05	100,000	999,999
06	1,000,000	9,999,999
07	10,000,000	49,999,999
08	50,000,000	99,999,999
09	100,000,000	499,999,999
10	500,000,000	999,999,999
11	1,000,000,000	>1,000,000,000

Table 10–3 SARA inventory storage types

Codes	Types of storage
A	Above ground tank
B	Below ground tank
C	Tank inside building
D	Steel drum
E	Plastic or non-metallic drum
F	Can
G	Carboy
H	Silo
I	Fiber drum
J	Bag
K	Box
L	Cylinder
M	Glass bottles or jugs
N	Plastic bottles or jugs
O	Tote bin
P	Tank wagon
Q	Rail car
R	Other

If the operator did not bring onsite 10,000 lb [500 lb in Louisiana] during the previous year, the inventory need not reflect that hazardous chemical even though a Section 311 notification previously was submitted. However, Louisiana will ask for a clarification that the inventory quantity equaled zero.

Table 10–4 SARA inventory temperature and pressure conditions

Codes	Storage conditions
	(Pressure)
1	Ambient pressure
2	Greater than ambient pressure
3	Less than ambient pressure
	(Temperature)
4	Ambient temperature
5	Greater than ambient temperature
6	Less than ambient temperature but not cryogenic
7	Cryogenic conditions

Chemical Release Reporting

The EPA[21] requires manufacturers [SIC codes 20-39] employing 10 or more persons and manufacturing or processing 25,000 lb or more of toxic chemicals[22] per year and/or 10,000 lb of a hazardous chemical per year to submit to the EPA and the state a completed EPA Form 9350-1[23] annually. *This provision currently does not impact E&P operators.* However, EPA (1995) reports:

> "Changes are also being studied by EPA which address the expansion of the SIC Code base to be used for submission of Form R [Form 9350-1] Reports. A Notice of Proposed Rule-making is now scheduled for Spring 1995"[24]

[21] 40 CFR Part 372
[22] 40 CFR §372.65
[23] 40 CFR §372.30
[24] EPA Small Business Ombudsman, Jan. 1995.

The Small Business Administration[25] warned the EPA of the high cost and inefficiency of the Section 313 reporting. This warning is particularly important for E&P operators, because produced water injected for disposal could be reported as a Section 313 release. The analysis of these waters for benzene [a toxic chemical] would be very expensive, time consuming, and wasteful since in reality the public is not exposed to any of the chemical. The American Petroleum Institute is working on this problem with the EPA.

Transportation Exemption

SARA Section 327 provides a useful exemption from the Section 311 notification and Section 312 inventory requirements. Section 327 exempts from Sections 311 and 312 all hazardous chemicals on a lease or facility when covered by active DOT shipping papers. This means the operator need not include in the SARA reporting any hazardous chemicals brought onto the lease by a chemical supplier while making a delivery or while batch treating wells.

This exemption does not apply to shipments which terminate at the lease, i.e., bulk cement, stimulation acid, or nitrogen shipments. The shipping papers terminate for these shipments when the chemical is delivered to the job site.

---

[25] 57 FR 48706

APPENDICES

I Hazardous Air Pollutants

CAS number	Chemical name
75070	Acetaldehyde
60355	Acetamide
75058	Acetonitrile
98862	Acetophenone
53963	2-Acetylaminofluorene
107028	Acrolein
79061	Acrylamide
79107	Acrylic acid
107131	Acrylonitrile
107051	Allyl chloride
92671	4-Aminobiphenyl
62533	Aniline
90040	o-Anisidine
1332214	Asbestos
71432	Benzene [including benzene from gasoline]
92875	Benzidine
98077	Benzotrichloride
100447	Benzyl chloride
92524	Biphenyl

117817	Bis(2-ethylhexyl)phthalate, [DEHP]
542881	Bis(chloromethyl)ether
75252	Bromoform
106990	1,3-Butadiene
156627	Calcium cyanamide
105602	Caprolactam
133062	Captan
63252	Carbaryl
75150	Carbon disulfide
56235	Carbon tetrachloride
463581	Carbonyl sulfide
120809	Catechol
133904	Chloramben
57749	Chlordane
7782505	Chlorine
79118	Chloroacetic acid
532274	2-Chloroacetophenone
108907	Chlorobenzene
510156	Chlorobenzilate
67663	Chloroform
107302	Chloromethyl methyl ether
126998	Chloroprene
1319773	Cresols/Cresylic acid [isomers and mixture]
95487	o-Cresol
108394	m-Cresol
106445	p-Cresol
98828	Cumene
94757	2,4-D, salts and esters
3547044	DDE
334883	Diazomethane
132649	Dibenzofurans
96128	1,2-Dibromo-3-chloropropane
84742	Dibutylphthalate
106467	1,4-Dichlorobenzene(p)
91941	3,3-Dichlorobenzidene
111444	Dichloroethyl ether [Bis(2-chloroethyl)ether]

542756	1,3-Dichloropropene
62737	Dichlorvos
111422	Diethanolamine
121697	N,N-Diethyl aniline [N,N-Dimethylaniline]
64675	Diethyl sulfate
119904	3,3-Dimethoxybenzidine
60117	Dimethyl aminoazobenzene
119937	3,3 -Dimethyl benzidine
79447	Dimethyl carbamoyl chloride
68122	Dimethyl formamide
57147	1,1-Dimethyl hydrazine
131113	Dimethyl phthalate
77781	Dimethyl sulfate
534521	4,6-Dinitro-o-cresol, and salts
51285	2,4-Dinitrophenol
121142	2,4-Dinitrotoluene
123911	1,4-Dioxane [1,4-Diethyleneoxide]
122667	1,2-Diphenylhydrazine
106898	Epichlorohydrin [l-Chloro-2,3-epoxypropane]
106887	1,2-Epoxybutane
140885	Ethyl acrylate
100414	Ethyl benzene
51796	Ethyl carbamate [Urethane]
75003	Ethyl chloride [Chloroethane]
106934	Ethylene dibromide [Dibromoethane]
107062	Ethylene dichloride [1,2-Dichloroethane]
107211	Ethylene glycol
151564	Ethylene imine [Aziridine]
75218	Ethylene oxide
96457	Ethylene thiourea
75343	Ethylidene dichloride [1,1-Dichloroethane]
50000	Formaldehyde
76448	Heptachlor
118741	Hexachlorobenzene

87683	Hexachlorobutadiene
77474	Hexachlorocyclopentadiene
67721	Hexachloroethane
822060	Hexamethylene-1,6-diisocyanate
680319	Hexamethylphosphoramide
110543	Hexane
302012	Hydrazine
7647010	Hydrochloric acid
7664393	Hydrogen fluoride [Hydrofluoric acid]
7783064	Hydrogen sulfide
123319	Hydroquinone
78591	Isophorone
58899	Lindane [all isomers]
108316	Maleic anhydride
67561	Methanol
72435	Methoxychlor
74839	Methyl bromide [Bromomethane]
74873	Methyl chloride [Chloromethane]
71556	Methyl chloroform [1,1,1-Trichloroethane]
78933	Methyl ethyl ketone [2-Butanone]
60344	Methyl hydrazine
74884	Methyl iodide [Iodomethane]
108101	Methyl isobutyl ketone [Hexone)
624839	Methyl isocyanate
80626	Methyl methacrylate
1634044	Methyl tert butyl ether
101144	4,4-Methylene bis [2-chloroaniline]
75092	Methylene chloride [Dichloromethane]
101688	Methylene diphenyl diisocyanate [MDI]
101779	4,4 -Methylenedianiline
91203	Naphthalene
98953	Nitrobenzene
92933	4-Nitrobiphenyl
100027	4-Nitrophenol

79469	2-Nitropropane
684935	N-Nitroso-N-methylurea
62759	N-Nitrosodimethylamine
59892	N-Nitrosomorpholine
56382	Parathion
82688	Pentachloronitrobenzene [Quintobenzene]
87865	Pentachlorophenol
108952	Phenol
106503	p-Phenylenediamine
75445	Phosgene
7803512	Phosphine
7723140	Phosphorus
85449	Phthalic anhydride
1336363	Polychlorinated biphenyls [Aroclors]
1120714	1,3-Propane sultone
57578	beta-Propiolactone
123386	Propionaldehyde
114261	Propoxur [Baygon]
78875	Propylene dichloride [1,2-Dichloropropane]
75569	Propylene oxide
75558	1,2-Propylenimine [2-Methyl aziridine]
91225	Quinoline
106514	Quinone
100425	Styrene
96093	Styrene oxide
1746016	2,3,7,8-Tetrachlorodibenzo-p-dioxin
79345	1,1,2,2-Tetrachloroethane
127184	Tetrachloroethylene [Perchloroethylene]
7550450	Titanium tetrachloride
108883	Toluene
95807	2,4-Toluene diamine
584849	2,4-Toluene diisocyanate

95534	o-Toluidine
8001352	Toxaphene [chlorinated camphene]
120821	1,2,4-Trichlorobenzene
79005	1,1,2-Trichloroethane
79016	Trichloroethylene
95954	2,4,5-Trichlorophenol
88062	2,4,6-Trichlorophenol
121448	Triethylamine
1582098	Trifluralin
540841	2,2,4-Trimethylpentane
108054	Vinyl acetate
593602	Vinyl bromide
75014	Vinyl chloride
75354	Vinylidene chloride [1,1-Dichloroethylene]
1330207	Xylenes [somers and mixture]
95476	o-Xylenes
108383	m-Xylenes
106423	p-Xylenes
0	Antimony Compounds
	Arsenic Compounds [inorganic including arsine]
0	Beryllium Compounds
0	Cadmium Compounds
0	Chromium Compounds
0	Cobalt Compounds
0	Coke Oven Emissions
0	Cyanide Compounds [1]
0	Glycol ethers [2]
0	Lead Compounds
0	Manganese Compounds
0	Mercury Compounds

0	Fine mineral fibers [3]
O	Nickel Compounds
	Polycylic Organic Matter [4]
	Radionuclides [including radon] [5]
0	Selenium Compounds

NOTE: For all listings above which contain the word *compounds* and for glycol ethers, the following applies: Unless otherwise specified, these listings are defined as including any unique chemical substance that contains the named chemical, i.e., antimony, arsenic, etc., as part of that chemical's infrastructure.

[1] X'CN where X = any group where a formal dissociation may occur. For example KCN or $Ca(CN)_2$

[2] Includes mono- and di- ethers of ethylene glycol, diethylene glycol, and triethylene glycol R-(OCH2CH2)$^{n-}$-OR' where

n = 1, 2, or 3
R - alkyl or aryl groups
R' = R. H, or groups which, when removed, yield Glyco ethers with the structure:
R- (OCH2CH)$_n$-OH. Polymers are excluded from the glycol category.

[3] Includes mineral fiber emissions from facilities manufacturing or processing glass, rock, or slag fibers (or other mineral derived fibers) of average diameter 1 micrometer or less.

[4] Includes organic compounds with more than one benzene ring, and which have a boiling point ≥100 C.

[5] A type of atom which spontaneously undergoes radioactive decay.

II Toxic Pollutants

1. Acenaphthene
2. Acrolein
3. Acrylonitrile
4. Aldrin/Dieldrin[1]
5. Antimony and compounds
6. Arsenic and compounds
7. Asbestos
8. Benzene
9. Benzidine[1]
10. Beryllium and compounds
11. Cadmium and com pounds
12. Carbon tetrachloride
13. Chlordane [technical mixture and metabolites]
14. Chlorinated benzenes [other than di-chloroben zenes]
15. Chlorinated ethanes [including 1,2-di-chloroethane, 1,1,1-trichloroethane, and hexachloroethane]
16. Chloroalkyl ethers [chloroethyl and mixed ethers]
17. Chlorinated naphthalene
18. Chlorinated phenols [other than those listed elsewhere; includes trichlorophenols and chlorinated cresols]
19. Chloroform
20. 2-chlorophenol
21. Chromium and com pounds
22. Copper and compounds
23. Cyanides
24. DDT and metabolites[1]

---

[1] Effluent standard promulgated (40 CFR Part 129)

[2] The term compounds shall include organic and inorganic compounds.

25. Dichlorobenzenes [1,2-, 1,3-, and 1,4-di-chlo robenzenes]
26. Dichlorobenzidine
27. Dichloroethylenes [1,1-, and 1,2-dichloroethylene]
28. 2,4-dichlorophenol
29. Dichloropropane and dichloropropene
30. 2,4-dimethylphenol
31. Dinitrotoluene
32. Diphenylhydrazine
33. Endosulfan and metabolites
34. Endrin and metabolites[1]
35. Ethylbenzene
36. Fluoranthene
37. Haloethers [other than those listed elsewhere; includes chlorophenylphenyl ethers, bromophenylphenyl ether, bis(dichloroisopropyl] ether, bis-(chloroethoxy) methane and polychlorinated diphenyl ethers]
38. Halomethanes (other than those listed elsewhere; includes methylene chloride, methylchloride, methylbromide, bromoform dichlorobromomethane
39. Heptachlor and metabo= lites
40. Hexachlorobutadiene
41. Hexachlorocyclohexane
42. Hexachlor ocyclopentadiene
43. Isophorone
44. Lead and compounds
45. Mercury and compounds
46. Naphthalene
47. Nickel and compounds
48. Nitrobenzene
49. Nitrophenols [including 2,4-dinitrophenol, dinitrocresol]
50. Nitrosamines
51. Pentachlorophenol
52. Phenol
53. Phthalate esters
54. Polychlorinated biphenyls [PCBs][1]
55. Polynuclear aromatic hydrocarbons [including benzanthracenes, benzopyrenes, benzofluoranthene, chrysenes, dibenz-an thracenes, and indenopyrenes]
56. Selenium and compounds
57. Silver and compounds
58. 2,3,7,8-tetrachlorod-ibenzo-p-dioxin [TCDD]
59. Tetrachloroethylene
60. Thallium and compounds
61. Toluene
62. Toxaphene1
63. Trichloroethylene
64. Vinyl chloride
65. Zinc and compounds

III Priority Pollutants

001 Acenaphthene
002 Acrolein
003 Acrylonitrile
004 Benzene
005 Benzidine
006 Carbon tetrachloride [tetrachloromethane]
007 Chlorobenzene
008 1,2,4-trichlorobenzene
009 Hexachlorobenzene
010 1,2-dichloroethane
011 1,1,1-trichloreothane
012 Hexachloroethane
013 1,1-dichloroethane
014 1,1,2-trichloroethane
015 1,1,2,2-tetrachloroethane
016 Chloroethane
018 Bis [2-chloroethyl] ether
019 2-chloroethyl vinyl ether [mixed]
020 2-chloronaphthalene
021 2,4, 6-trichlorophenol
022 Parachlorometa cresol
023 Chloroform [trichloromethane[
024 2-chlorophenol
025 1,2-dichlorobenzene
026 1,3-dichlorobenzene
027 1,4-dichlorobenzene
028 3,3-dichlorobenzidine
029 1,1-dichloroethylene
030 1,2-trans-dichloroethylene
031 2,4-dichlorophenol
032 1,2-dichloropropane
033 1,2-dichloropropylene [1,3-dichloropropene]
034 2,4-dimethylphenol
035 2,4-dinitrotoluene
036 2,6-dinitrotoluene
037 1,2-diphenylhydrazine
038 Ethylbenzene
039 Fluoranthene
040 4-chlorophenyl phenyl ether
041 4-bromophenyl phenyl ether

042 Bis [2-chloroisopropyl] ether
043 Bis [2-chloroethoxy] meth ane
044 Methylene chloride [dichloromethane]
045 Methyl chloride [dichloromethane]
046 Methyl bromide [bromomethane]
047 Bromoform [tribromomethane]
048 Dichlorobromomethane
051 Chlorodibromomethane
052 Hexachlorobutadiene
053 Hexachloromyclopentadiene
054 Isophorone
055 Naphthalene
056 Nitrobenzene
057 2-nitrophenol
058 4-nitrophenol
059 2,4-dinitrophenol
060 4,6-dinitro-o-cresol
061 N-nitrosodimethylamine
062 N-nitrosodiphenylamine
063 N-nitrosodi-n-propylamin
064 Pentachlorophenol
065 Phenol
066 Bis[2-ethylhexyl] phthalate
067 Butyl benzyl phthalate
068 Di-N-Butyl Phthalate
069 Di-n-octyl phthalate
070 Diethyl Phthalate
071 Dimethyl phthalate
072 1,2-benzanthracene [benzo(a) anthracene]
073 Benzo(a)pyrene [3,4-benzo-pyrene]
074 3,4-Benzofluoranthene [benzo(b) fluoranthene]
075 11,12-benzofluoranthene [benzo(b) fluoranthene]
076 Chrysene
077 Acenaphthylene
078 Anthracene
079 1,12-benzoperylene [benzo(ghi) perylene]
080 Fluorene
081 Phenanthrene
082 1,2,5,6-dibenzanthracene [dibenzo(,h) anthracene]
083 Indeno [,1,2,3-cd] pyrene [2,3-o-pheynylene pyrene]
084 Pyrene
085 Tetrachloroethylene
086 Toluene
087 Trichloroethylene
088 Vinyl chloride [chloroethylene]
089 Aldrin
090 Dieldrin
091 Chlordane [technical mixture and metabolites]
092 4,4-DDT
093 4,4-DDE [p,p-DDX]
094 4,4-DDD [p,p-TDE]
095 Alpha-endosulfan
096 Beta-endosulfan
097 Endosulfan sulfate
098 Endrin
099 Endrin aldehyde
100 Heptachlor

101 Heptachlor epoxide [BHC-hexachlorocyclohexane]
102 Alpha-BHC
103 Beta-BHC
104 Gamma-BHC (lindane)
105 Delta-BHC (PCB-polychlor-inated biphenyls)
106 PCB-1242 [Arochlor 1242]
107 PCB-1254 [Arochlor 1254]
108 PCB-1221 [Arochlor 1221]
109 PCB-1232 [Arochlor 1232]
110 PCB-1248 [Arochlor 1248]
111 PCB-1260 [Arochlor 1260]
112 PCB-1016 [Arochlor 1016]
113 Toxaphene
114 Antimony
115 Arsenic
116 Asbestos
117 Beryllium
118 Cadmium
119 Chromium
120 Copper
121 Cyanide, Total
122 Lead
123 Mercury
124 Nickel
125 Selenium
126 Silver
127 Thallium
126 Silver
128 Zinc
129 2,3,7,8-tetrachloro-dibenzo-p-dioxin [TCDD]

IV List of Hazardous Substances and Reportable Quantities

Note: All comments/notes are located at the end of this table.

			Statutory			Final RQ	
Hazardous substance	CASRN	Regulatory synonyms	RQ	Code	RCRA waste Number	Category	Pounds (Kg)
Acenaphthene	83329		1*	2		B	100 (45.4)
Acenaphthylene	208968		1*	2		D	5000 (2270)
Acetaldehyde	75070	Ethanal	1000	1,4	U001	C	1000 (454)
Acetaldehyde, chloro-	107200	Chloroace taldehyde	1*	4	P023	C	1000 (454)
Acetaldehyde, trichloro-	75876	Chloral	1*	4	U034	D	5000 (2270)
Acetamide, N-(aminothioxom ethyl)-	591082	1-Acetyl-2- -thiourea	1*	4	P002	C	1000 (454)
Acetamide, N-(4-ethoxyphen yl)-	62442	Phenaceti n	1*	4	U187	B	100 (45.4)
Acetamide, 2-fluoro-	640197	Fluoroacet amide	1*	4	P057	B	100 (45.4)
Acetamide, N-9H-fluoren-2-yl- -	53963	2-Acetyla minofluore ne	1*	4	U005	X	1 (0.454)
Acetic acid	64197		1000	1		D	5000 (2270)
Acetic acid (2,4-dichlorophen oxy)-	94757	2,4-D Acid	100	1,4	U240	B	100 (45.4)
		2,4-D, salts and esters					
Acetic acid, Lead(2+) salt	301042	Lead acetate	5000	1,4	U144	A	10 (4.54)
Acetic acid, thallium (1+) salt	563688	Thallium(I) acetate	1*	4	U214	B	100 (45.4)
Acetic acid, (2,4,5-trichloroph enoxy)	93765	2,4,5-T<cr>2,4,5-T acid	100	1,4	U232	C	1000 (454)
Acetic acid, ethyl ester	141786	Ethyl acetate	1*	4	U112	D	5000 (2270)
Acetic acid, fluoro-, sodium salt	62748	Fluoroacet ic acid, sodium salt	1*	4	P058	A	10 (4.54)
Acetic anhydride	108247		1000	1		D	5000 (2270)
Acetone	67641	2-Propano ne	1*	4	U002	D	5000 (2270)
Acetone cyanohydrin	75865	Propaneni trile, 2-hydroxy- -2-methyl- 2-Methylla ctonitrile	10	1,4	P069	A	10 (4.54)

Acetonitrile	75058		1*	4	U003	D	5000 (2270)
Acetophenone	98862	Ethanone, 1-phenyl-	1*	4	U004	D	5000 (2270)
2-Acetylaminofluorene	53963	Acetamide, N-9H-fluoren-2-yl-	1*	4	U005	X	1 (0.454)
Acetyl bromide	506967		5000	1		D	5000 (2270)
Acetyl chloride	75365		5000	1,4	U006	D	5000 (2270)
1-Acetyl-2-thiourea	591082	Acetamide, N-(aminothioxomethyl)-	1*	4	P002	C	1000 (454)
Acrolein	107028	2-Propenal	1	1,2,4	P003	X	1 (0.454)
Acrylamide	79061	2-Propenamide	1*	4	U007	D	5000 (2270)
Acrylic acid	79107	2-Propenoic acid	1*	4	U008	D	5000 (2270)
Acrylonitrile	107131	2-Propenenitrile	100	1,2,4	U009	B	100 (45.4)
Adipic acid	124049		5000	1		D	5000 (2270)
Aldicarb	116063	Propanal, 2-methyl-2-(methylthio)-,O-[(methylamino)carbonyl]oxime	1*	4	P070	X	1 (0.454)
Aldrin	309002	1,4,5,8-Dimethanonaphthalene, 1,2,3,4,10,10-10-hexachloro-1,4,4a,5,8,8a-hexahydro-, (1alpha,4alpha,4abeta,5alpha,8alpha,8abeta)-	1	1,2,4	P004	X	1 (0.454)
Allyl alcohol	107186	2-Propen-1-ol	100	1,4	P005	B	100 (45.4)
Allyl chloride	107051		1000	1		C	1000 (454)
Aluminum phosphide	20859738		1*	4	P006	B	100 (45.4)
Aluminum sulfate	10043013		5000	1		D	5000 (2270)
5-(Aminomethyl)-3-isoxazolol	2763964	Muscimol 3(2H)-Isoxazolone, 5-(aminomethyl)-	1*	4	P007	C	1000 (454)
4-Aminopyridine	504245	4-Pyridinamine	1*	4	P008	C	1000 (454)

Amitrole	61825	1H-1,2,4-Triazol-3-amine	1*	4	U011	A	10 (4.54)
Ammonia	7664417		100	1		B	100 (45.4)
Ammonium acetate	631618		5000	1		D	5000 (2270)
Ammonium benzoate	1863634		5000	1		D	5000 (2270)
Ammonium bicarbonate	1066337		5000	1		D	5000 (2270)
Ammonium bichromate	7789095		1000	1		A	10 (4.54)
Ammonium bifluoride	1341497		5000	1		B	100 (45.4)
Ammonium bisulfite	10192300		5000	1		D	5000 (2270)
Ammonium carbamate	1111780		5000	1		D	5000 (2270)
Ammonium carbonate	506876		5000	1		D	5000 (2270)
Ammonium chloride	12125029		5000	1		D	5000 (2270)
Ammonium chromate	7788989		1000	1		A	10 (4.54)
Ammonium citrate, dibasic	3012655		5000	1		D	5000 (2270)
Ammonium fluoborate	13826830		5000	1		D	5000 (2270)
Ammonium fluoride	12125018		5000	1		B	100 (45.4)
Ammonium hydroxide	1336216		1000	1		C	1000 (454)
Ammonium oxalate	6009707<cr>5972736<cr>14258492		5000	1		D	5000 (2270)
Ammonium picrate	131748	Phenol, 2,4,6-trinitro-, ammonium salt	1*	4	P009	A	10 (4.54)
Ammonium silicofluoride	16919190		1000	1		C	1000 (454)
Ammonium sulfamate	7773060		5000	1		D	5000 (2270)
Ammonium sulfide	12135761		5000	1		B	100 (45.4)
Ammonium sulfite	10196040		5000	1		D	5000 (2270)
Ammonium tartrate	14307438<cr>3164292		5000	1		D	5000 (2270)
Ammonium thiocyanate	1762954		5000	1		D	5000 (2270)
Ammonium vanadate	7803556	Vanadic acid, ammonium salt	1*	4	P119	C	1000 (454)
Amyl acetate	628637		1000	1		D	5000 (2270)
iso-Amyl acetate	123922						

sec-Amyl acetate	626380						
tert-Amyl acetate	625161						
Aniline	62533	Benzenamine	1000	1,4	U012	D	5000 (2270)
Anthracene	120127		1*	2		D	5000 (2270)
Antimony	7440360		1*	2		D	5000 (2270)
ANTIMONY AND COMPOUNDS	N.A.		1*	2			**
Antimony pentachloride	7647189		1000	1		C	1000 (454)
Antimony potassium tartrate	28300745		1000	1		B	100 (45.4)
Antimony tribromide	7789619		1000	1		C	1000 (454)
Antimony trichloride	10025919		1000	1		C	1000 (454)
Antimony trifluoride	7783564		1000	1		C	1000 (454)
Antimony trioxide	1309644		5000	1		C	1000 (454)
Argentate(1-), bis(cyano-C)-, potassium	506616	Potassium silver cyanide	1*	4	P099	X	1 (0.454)
Aroclor 1016	12674112	POLYCHLORINATED BIPHENYLS (PCBs)	10	1,2		X	1 (0.454)
Aroclor 1221	11104282	POLYCHLORINATED BIPHENYLS (PCBs)	10	1,2		X	1 (0.454)
Aroclor 1232	11141165	POLYCHLORINATED BIPHENYLS (PCBs)	10	1,2		X	1 (0.454)
Aroclor 1242	53469219	POLYCHLORINATED BIPHENYLS (PCBs)	10	1,2		X	1 (0.454)
Aroclor 1248	12672296	POLYCHLORINATED BIPHENYLS (PCBs)	10	1,2		X	1 (0.454)
Aroclor 1254	11097691	POLYCHLORINATED BIPHENYLS (PCBs)	10	1,2		X	1 (0.454)
Aroclor 1260	11096825	POLYCHLORINATED BIPHENYLS (PCBs)	10	1,2		X	1 (0.454)

Arsenic	7440382		1*	2,3		X	1 (0.454)
Arsenic acid	1327522 <cr>7778394	Arsenic acid H3AsO4	1*	4	P010	X	1 (0.454)
Arsenic acid H3AsO4	1327522 <cr>7778394	Arsenic acid	1*	4	P010	X	1 (0.454)
ARSENIC AND COMPOUNDS	N.A.		1*	2			**
Arsenic disulfide	1303328		5000	1		X	1 (0.454)
Arsenic oxide As2O3	1327533	Arsenic trioxide	5000	1,4	P012	X	1 (0.454)
Arsenic oxide As2O5	1303282	Arsenic pentoxide	5000	1,4	P011	X	1 (0.454)
Arsenic pentoxide	1303282	Arsenic oxide As2O5	5000	1,4	P011	X	1 (0.454)
Arsenic trichloride	7784341		5000	1		X	1 (0.454)
Arsenic trioxide	1327533	Arsenic oxide As2O3	5000	1,4	P012	X	1 (0.454)
Arsenic trisulfide	1303339		5000	1		X	1 (0.454)
Arsine, diethyl-	692422	Diethylarsine	1*	4	P038	X	1 (0.454)
Arsinic acid, dimethyl-	75605	Cacodylic acid	1*	4	U136	X	1 (0.454)
Arsonous dichloride, phenyl-	696286	Dichlorophenylarsine	1*	4	P036	X	1 (0.454)
Asbestos	1332214		1*	2,3		X	1 (0.454)
Auramine	492808	Benzenamine, 4,4'-carbonimidoylbis (N,N-dimethyl-	1*	4	U014	B	100 (45.4)
Azaserine	115026	L-Serine, diazoacetate (ester)	1*	4	U015	X	1 (0.454)
Aziridine	151564	Ethylenimine	1*	4	P054	X	1 (0.454)
Aziridine, 2-methyl-	75558	1,2-Propylenimine	1*	4	P067	X	1 (0.454)
Azirino[2',3':3,4]pyrrolo[1,2-a]indole-4,7-dione,6-amino-8-[[(aminocarbonylooxy]methyl]-1,1a,2,8,8a,8b-hexahydro-8a--methoxy-5-methyl-,[1aS-(1aalpha,8beta,8aalpha,8balpha)]-	50077	Mitomycin C	1*	4	U010	A	10 (4.54)
Barium cyanide	542621		10	1,4	P013	A	10 (4.54)
Benz[j]aceanthrylene, 1,2-dihydro-3-methyl-	56495	3-Methylcholanthrene	1*	4	U157	A	10 (4.54)
Benz[c]acridine	225514		1*	4	U016	B	100 (45.4)

Benzal chloride	98873	Benzene, dichloromethyl-	1*	4	U017	D	5000 (2270)
Benzamide, 3,5-dichloro-N-(1,1-dimethyl-2-propynyl)-	23950585	Pronamide	1*	4	U192	D	5000 (2270)
Benz[a]anthracene	56553	Benzo[a]anthracene<cr>1,2-Benzanthracene	1*	2,4	U018	A	10 (4.54)
1,2-Benzanthracene	56553	Benz[a]anthracene<cr>Benzo[a]anthracene	1*	2,4	U018	A	10 (4.54)
Benz[a]anthracene, 7,12-dimethyl-	57976	7,12-Dimethylbenz[a]anthracene	1*	4	U094	X	1 (0.454)
Benzenamine	62533	Aniline	1000	1,4	U012	D	5000 (2270)
Benzenamine, 4,4'-carbonimidoylbis (N,N-dimethyl-	492808	Auramine	1*	4	U014	B	100 (45.4)
Benzenamine, 4-chloro-	106478	p-Chloroaniline	1*	4	P024	C	1000 (454)
Benzenamine, 4-chloro-2-methyl--, hydrochloride	3165933	4-Chloro-o-toluidine, hydrochloride	1*	4	U049	B	100 (45.4)
Benzenamine, N,N-dimethyl-4-(phenylazo-)	60117	p-Dimethylaminoazobenzene	1*	4	U093	A	10 (4.54)
Benzenamine, 2-methyl-	95534	o-Toluidine	1*	4	U328	B	100 (45.4)
Benzenamine, 4-methyl-	106490	p-Toluidine	1*	4	U353	B	100 (45.4)
Benzenamine, 4,4'-methylenebis (2-chloro-	101144	4,4'-Methylenebis(2--chloroaniline)	1*	4	U158	A	10 (4.54)
Benzenamine, 2-methyl-, hydrochloride	636215	o-Toluidine hydrochloride	1*	4	U222	B	100 (45.4)
Benzenamine, 2-methyl-5-nitro-	99558	5-Nitro-o-toluidine	1*	4	U181	B	100 (45.4)
Benzenamine, 4-nitro-	100016	p-Nitroaniline	1*	4	P077	D	5000 (2270)
Benzene	71432		1000	1,2,3,4	U109	A	10 (4.54)
Benzeneacetic acid, 4-chloro-alpha-(4--chlorophenyl)-alpha-hydroxy-, ethyl ester	510156	Chlorobenzilate	1*	4	U038	A	10 (4.54)
Benzene, 1-bromo-4-phenoxy-	101553	4-Bromophenyl phenyl ether	1*	2,4	U030	B	100 (45.4)

Benzenebutanoic acid, 4-[bis(2-chloroethyl)amino]-	305033	Chlorambucil	1*	4	U035	A	10 (4.54)
Benzene, chloro-	108907	Chlorobenzene	100	1,2,4	U037	B	100 (45.4)
Benzene, chloromethyl-	100447	Benzyl chloride	100	1,4	P028	B	100 (45.4)
Benzenediamin, ar-methyl-	95807 <cr>496720 <cr>823405	Toluenediamine	1*	4	U221	A	10 (4.54)
1,2-Benzenedicarboxylic acid, dioctyl ester	117840	Di-n-octyl phthalate	1*	2,4	U107	D	5000 (2270)
1,2-Benzenedicarboxylic acid, [bis(2-ethylhexyl)]--ester	117817	Bis (2-ethylhexyl)phthalate<cr>Diethylhexyl phthalate	1*	2,4	U028	B	100 (45.4)
1,2-Benzenedicarboxylic acid, dibutyl ester	84742	Di-n-butyl phthalate <cr>Dibutyl phthalate<cr>n-Butyl phthalate	100	1,2,4	U069	A	10 (4.54)
1,2-Benzenedicarboxylic acid, diethyl ester	84662	Diethyl phthalate	1*	2,4	U088	C	1000 (454)
1,2-Benzenedicarboxylic acid, dimethyl ester	131113	Dimethyl phthalate	1*	2,4	U102	D	5000 (2270)
Benzene, 1,2-dichloro-	95501	o-Dichlorobenzene<cr>1,2-Dichlorobenzene	100	1,2,4	U070	B	100 (45.4)
Benzene, 1,3-dichloro-	541731	m-Dichlorobenzene <cr>1,3-Dichlorobenzene	1*	2,4	U071	B	100 (45.4)
Benzene, 1,4-dichloro-	106467	p-Dichlorobenzene <cr>1,4-Dichlorobenzene	100	1,2,4	U072	B	100 (45.4)
Benzene, 1,1'-(2,2-dichloroethylidene)bis[4-chloro-	72548	DDD <cr>TDE <cr>4,4' DDD	1	1,2,4	U060	X	1 (0.454)
Benzene, dichloromethyl-	98873	Benzal chloride	1*	4	U017	D	5000 (2270)
Benzene, 1,3-diisocyanatomethyl-	584849 <cr>91087 <cr>26471625	Toluene diisocyanate	1*	4	U223	B	100 (45.4)
Benzene, dimethyl	1330207	Xylene (mixed)	1000	1,4	U239	C	1000 (454)

m-Benzene, dimethyl	108383	m-Xylene					
o-Benzene, dimethyl	95476	o-Xylene					
p-Benzene, dimethyl	106423	p-Xylene					
1,3-Benzenediol	108463	Resorcinol	1000	1,4	U201	D	5000 (2270)
1,2-Benzenediol, 4-[1-hydroxy-2-(methylamino)ethyl]-	51434	Epinephrine	1*	4	P042	C	1000 (454)
Benzeneethanamine, alpha,alpha-dimethyl-	122098	alpha,alpha-Dimethylphenethylamine	1*	4	P046	D	5000 (2270)
Benzene, hexachloro-	118741	Hexachlorobenzene	1*	2,4	U127	A	10 (4.54)
Benzene, hexahydro-	110827	Cyclohexane	1000	1,4	U056	C	1000 (454)
Benzene, hydroxy-	108952	Phenol	1000	1,2,4	U188	C	1000 (454)
Benzene, methyl-	108883	Toluene	1000	1,2,4	U220	C	1000 (454)
Benzene, 2-methyl-1,3-dinitro-	606202	2,6-Dinitrotoluene	1000	1,2,4	U106	B	100 (45.4)
Benzene, 1-methyl-2,4-dinitro-	121142	2,4-Dinitrotoluene	1000	1,2,4	U105	A	10 (4.54)
Benzene, 1-methylethyl-	98828	Cumene	1*	4	U055	D	5000 (2270)
Benzene, nitro-	98953	Nitrobenzene	1000	1,2,4	U169	C	1000 (454)
Benzene, pentachloro-	608935	Pentachlorobenzene	1*	4	U183	A	10 (4.54)
Benzene, pentachloronitro-	82688	Pentachloronitrobenzene (PCNB)	1*	4	U185	B	100 (45.4)
Benzenesulfonic acid chloride	98099	Benzenesulfonyl chloride	1*	4	U020	B	100 (45.4)
Benzenesulfonyl chloride	98099	Benzenesulfonic acid chloride	1*	4	U020	B	100 (45.4)
Benzene, 1,2,4,5-tetrachloro-	95943	1,2,4,5-Tetrachlorobenzene	1*	4	U207	D	5000 (2270)
Benzenethiol	108985	Thiophenol	1*	4	P014	B	100 (45.4)
Benzene, 1,1'-(2,2,2-tri-chloroethylidene) bis[4-chloro-	50293	DDT <cr>4,4'DDT	1	1,2,4	U061	X	1 (0.454)
Benzene, 1,1'-(2,2,2-trichloroethylidene) bis[4-methoxy-	72435	Methoxychlor	1	1,4	U247	X	1 (0.454)
Benzene, (trichloromethyl)-	98077	Benzotrichloride	1*	4	U023	A	10 (4.54)

Benzene, 1,3,5-trinitro-	99354	1,3,5-Trinitrobenzene	1*	4	U234	A	10 (4.54)
Benzidine	92875	(1,1'-Biphenyl)-4,4' diamine	1*	2,4	U021	X	1 (0.454)
1,2-Benzisothiazol-3(2H)-one, 1,1-dioxide	81072	Saccharin and salts	1*	4	U202	B	100 (45.4)
Benzo[a]anthracene	56553	Benz[a]anthracene<cr>1,2-Benzanthracene	1*	2,4	U018	A	10 (4.54)
Benzo[b]fluoranthene	205992		1*	2		X	1 (0.454)
Benzo(k)fluoranthene	207089		1*	2		D	5000 (2270)
Benzo[j,k]fluorene	206440	Fluoranthene	1*	2,4	U120	B	100 (45.4)
1,3-Benzodioxole, 5-)1-propenyl)-	120581	Isosafrole	1*	4	U141	B	100 (45.4)
1,3-Benzodioxole,5-(2-propenyl)-	94597	Safrole	1*	4	U203	B	100 (45.4)
1,3-Benzodioxole, 5-propyl-	94586	Dihydrosafrole	1*	4	U090	A	10 (4.54)
Benzoic acid	65850		5000	1		D	5000 (2270)
Benzonitrile	100470		1000	1		D	5000 (2270)
Benzo [rst]pentaphene	189559	Dibenz[a,i]pyrene	1*	4	U064	A	10 (4.54)
Benzo[ghi]perylene	191242		1*	2		D	5000 (2270)
2H-1-Benzopyran--2-one, 4-hydroxy-3-(3-oxo-1-phenyl-butyl)-, & salts, when present at concentrations greater than 0.3%	81812	Warfarin, & salts, when present at concentrations greater than 0.3%	1*	4	P001	B	100 (45.4)
Benzo[a]pyrene	50328	3,4-Benzopyrene	1*	2,4	U022	X	1 (0.454)
3,4-Benzopyrene	50328	Benzo[a]pyrene	1*	2,4	U022	X	1 (0.454)
p-Benzoquinone	106514	2,5-Cyclohexadiene--1,4-dione	1*	4	U197	A	10 (4.54)
Benzotrichloride	98077	Benzene, (trichloromethyl)-	1*	4	U023	A	10 (4.54)
Benzoyl chloride	98884		1000	1		C	1000 (454)
1,2-Benzphenanthrene	218019	Chrysene	1*	2,4	U050	B	100 (45.4)
Benzyl chloride	100447	Benzene, chloromethyl-	100	1,4	P028	B	100 (45.4)
BERYLLIUM AND COMPOUNDS	N.A.		1*	2			**
Beryllium chloride	7787475		5000	1		X	1 (0.454)
Beryllium fluoride	7787497		5000	1		X	1 (0.454)

Beryllium nitrate	13597994<cr>7787555		5000	1		X	1 (0.454)
Beryllium powder	7440417	Beryllium	1*	2,3,4	P015	A	10 (4.54)
alpha-BHC	319846		1*	2		A	10 (4.54)
beta-BHC	319857		1*	2		X	1 (0.454)
delta-BHC	319868		1*	2		X	1 (0.454)
gamma-BHC	58899	Cyclohexane, 1,2,3,4,5,6-hexachloro-,(1alpha, 2alpha,3beta,4alpha,5alpha, 6beta)-Hexachlorocyclohexane (gamma isomer) Lindane	1	1,2,4	U129	X	1 (0.454)
2,2'-Bioxirane	1464535	1,2:3,4-Diepoxybutane	1*	4	U085	A	10 (4.54)
(1,1'-Biphenyl)-4, 4'diamine	92875	Benzidine	1*	2,4	U021	X	1 (0.454)
[1,1'-Biphenyl]-4, 4'diamine,3,3'dichloro-	91941	3,3'-Dichlorobenzidine	1*	2,4	U073	X	1 (0.454)
[1,1'-Biphenyl]-4, 4'diamine,3,3'dimethoxy-	119904	3,3'-Dimethoxybenzidine	1*	4	U091	B	100 (45.4)
[1,1'Biphenyl]-4, 4'-diamine,3,3'-dimethyl-	119937	3,3'-Dimethylbenzidine	1*	4	U095	A	10 (4.54)
Bis (2-chloroethyl) ether	111444	Dichloroethyl ether <cr>Ethane,1,1'-oxybis[2-chloro-	1*	2,4	U025	A	10 (4.54)
Bis(2-chloroethoxy) methane	111911	Dichloromethoxy ethane <cr>Ethane, 1,1'-[methylenebis(oxy)]bis(2-chloro-	1*	2,4	U024	C	1000 (454)
Bis (2-ethylhexyl)phthalate	117817	Diethylhexyl phthalate <cr>1,2-Benzenedicarboxylic acid, [bis(2-ethylhexyl)] ester	1*	2,4	U028	B	100 (45.4)
Bromoacetone	598312	2-Propanone, 1-bromo-	1*	4	P017	C	1000 (454)

Bromoform	75252	Methane, tribromo-	1*	2,4	U225	B	100 (45.4)
4-Bromophenyl phenyl ether	101553	Benzene, 1-bromo-4-phenoxy-	1*	2,4	U030	B	100 (45.4)
Brucine	357573	Strychnidin-10-one, 2,3-dimethoxy-	1*	4	P018	B	100 (45.4)
1,3-Butadiene, 1,1,2,3,4,4-hexachloro-	87683	Hexachlorobutadiene	1*	2,4	U128	X	1 (0.454)
1-Butanamine, N-butyl-N-nitroso--	924163	N-Nitrosodi-n-butylamine	1*	4	U172	A	10 (4.54)
1-Butanol	71363	n-Butyl alcohol	1*	4	U031	D	5000 (2270)
2-Butanone	78933	Methyl ethyl ketone (MEK)	1*	4	U159	D	5000 (2270)
2-Butanone peroxide	1338234	Methyl ethyl ketone peroxide	1*	4	U160	A	10 (4.54)
2-Butanone, 3,3-dimethyl-1-(methylthio)-, O[(methylamino)carbonyl] oxime	39196184	Thiofanox	1*	4	P045	B	100 (45.4)
2-Butenal	123739 <cr>4170303	Crotonaldehyde	100	1,4	U053	B	100 (45.4)
2-Butene, 1,4-dichloro-	764410	1,4-Dichloro-2-butene	1*	4	U074	X	1 (0.454)
2-Butenoic acid, 2-methyl-, 7[[2,3-dihydroxy-2-(1-methoxyethyl)-3-methyl-1-oxobutoxy]methyl]-2,3,5,7a-tetrahydro-1H-pyrrolizin-1-yl ester, [1S-[1alpha(Z),7(2S*,3R*),7aalpha]]-	303344	Lasiocarpine	1*	4	U143	A	10 (4.54)
Butyl acetate	123864		5000	1		D	5000 (2270)
iso-Butyl acetate	110190						
sec-Butyl acetate	105464						
tert-Butyl acetate	540885						
n-Butyl alcohol	71363	1-Butanol	1*	4	U031	D	5000 (2270)
Butylamine	109739		1000	1		C	1000 (454)
iso-Butylamine	78819						
sec-Butylamine	513495 <cr>13952846						
tert-Butylamine	75649						

Butyl benzyl phthalate	85687		1*	2		B	100 (45.4)
n-Butyl phthalate	84742	Di-n-butyl phthalate <cr>Dibutyl phthalate <cr>1,2-Benzenedicarboxylic acid, dibutyl ester	100	1,2,4	U069	A	10 (4.54)
Butyric acid	107926		5000	1		D	5000 (2270)
iso-Butyric acid	79312						
Cacodylic acid	75605	Arsinic acid, dimethyl-	1*	4	U136	X	1 (0.454)
Cadmium	7440439		1*	2		A	10 (4.54)
Cadmium acetate	543908		100	1		A	10 (4.54)
CADMIUM AND COMPOUNDS	N.A.		1*	2			**
Cadmium bromide	7789426		100	1		A	10 (4.54)
Cadmium chloride	10108642		100	1		A	10 (4.54)
Calcium arsenate	7778441		1000	1		X	1 (0.454)
Calcium arsenite	52740166		1000	1		X	1 (0.454)
Calcium carbide	75207		5000	1		A	10 (4.54)
Calcium chromate	13765190	Chromic acid H2CrO4, calcium salt	1000	1,4	U032	A	10 (4.54)
Calcium cyanide	592018	Calcium cyanide Ca(CN)2	10	1,4	P021	A	10 (4.54)
Calcium cyanide Ca(CN)2	592018	Calcium cyanide	10	1,4	P021	A	10 (4.54)
Calcium dodecylbenzenesulfonate	26264062		1000	1		C	1000 (454)
Calcium hypochlorite	7778543		100	1		A	10 (4.54)
Camphene, octachloro-	8001352	Toxaphene	1	1,2,4	P123	X	1 (0.454)
Captan	133062		10	1		A	10 (4.54)
Carbamic acid, ethyl ester	51796	Ethyl carbamate (urethane)	1*	4	U238	B	100 (45.4)
Carbamic acid, methylnitroso-, ethyl ester	615532	N-Nitroso-N-methyl urethane	1*	4	U178	X	1 (0.454)
Carbamic chloride, dimethyl-	79447	Dimethylcarbamoyl chloride	1*	4	U097	X	1 (0.454)

Carbamodithioic acid, 1,2-ethanediylbis, salts & esters	111546	Ethylenebisdithiocarbamic acid, salts & esters	1*	4	U114	D	5000 (2270)
Carbamothioic acid, bis(1-methylethyl)--, S-(2,3-dichloro-2--propenyl) ester	2303164	Diallate	1*	4	U062	B	100 (45.4)
Carbaryl	63252		100	1		B	100 (45.4)
Carbofuran	1563662		10	1		A	10 (4.54)
Carbon disulfide	75150		5000	1,4	P022	B	100 (45.4)
Carbon oxyfluoride	353504	Carbonic difluoride	1*	4	U033	C	1000 (454)
Carbon tetrachloride	56235	Methane, tetrachloro--	5000	1,2,4	U211	A	10 (4.54)
Carbonic acid, dithallium(1+) salt	6533739	Thallium(I) carbonate	1*	4	U215	B	100 (45.4)
Carbonic dichloride	75445	Phosgene	5000	1,4	P095	A	10 (4.54)
Carbonic difluoride	353504	Carbon oxyfluoride	1*	4	U033	C	1000 (454)
Carbonochloridic acid, methyl ester	79221	Methyl chlorocarbonate<cr>Methyl chloroformate	1*	4	U156	C	1000 (454)
Chloral	75876	Acetaldehyde, trichloro-	1*	4	U034	D	5000 (2270)
Chlorambucil	305033	Benzenebutanoic acid, 4-[bis(2-chloroethyl) amino]-	1*	4	U035	A	10 (4.54)
Chlordane	57749	Chlordane, alpha & gamma isomers <cr>Chlordane, technical <cr>4,7-Methano-1H-indene, 1,2,4,5,6,7,8,8-octachloro-2,3,3a,4,7,7a-hexahydro-	1	1,2,4	U036	X	1 (0.454)
CHLORDANE (TECHNICAL MIXTURE AND METABOLITES)	N.A.		1*	2			**

Chlordane, alpha & gamma isomers	57749	Chlordane <cr>Chlordane, technical <cr>4,7-Methano-1H-indene, 1,2,4,5,6,7,8,8-octachloro-2,3,3a,4,7,7a-hexahydro-	1	1,2,4	U036	X	1 (0.454)
Chlordane, technical	57749	Chlordane <cr>Chlordane, alpha & gamma isomers <cr>4,7-Methano-1H-indene, 1,2,4,5,6,7,8,8-octachloro-2,3,3a,4,7,7a-hexahydro-	1	1,2,4	U036	X	1 (0.454)
CHLORINATED BENZENES	N.A.		1*	2			**
CHLORINATED ETHANES	N.A.		1*	2			**
CHLORINATED NAPHTHALENE	N.A.		1*	2			**
CHLORINATED PHENOLS	N.A.		1*	2			**
Chlorine	7782505		10	1		A	10 (4.54)
Chlornaphazine	494031	Naphthalenamine, N,N'-bis(2-chloroethyl)-	1*	4	U026	B	100 (45.4)
Chloroacetaldehyde	107200	Acetaldehyde, chloro-	1*	4	P023	C	1000 (454)
CHLOROALKYL ETHERS	N.A.		1*	2			**
p-Chloroaniline	106478	Benzenamine, 4-chloro-	1*	4	P024	C	1000 (454)
Chlorobenzene	108907	Benzene, chloro-	100	1,2,4	U037	B	100 (45.4)
Chlorobenzilate	510156	Benzeneacetic acid, 4-chloro-alpha-(4-chlorophenyl)-alpha-hydroxy-, ethyl ester	1*	4	U038	A	10 (4.54)

4-Chloro-m-cresol	59507	p-Chloro-m-cresol <cr>Phenol, 4-chloro-3--methyl-	1*	2,4	U039	D	5000 (2270)
p-Chloro-m-cresol	59507	Phenol, 4-chloro-3--methyl-<cr>4-Chloro-m-cresol	1*	2,4	U039	D	5000 (2270)
Chlorodibromomethane	124481		1*	2		B	100 (45.4)
Chloroethane	75003		1*	2		B	100 (45.4)
2-Chloroethyl vinyl ether	110758	Ethene, 2-chloroethoxy-	1*	2,4	U042	C	1000 (454)
Chloroform	67663	Methane, trichloro-	5000	1,2,4	U044	A	10 (4.54)
Chloromethyl methyl ether	107302	Methane, chloromethoxy-	1*	4	U046	A	10 (4.54)
beta-Chloronaphthalene	91587	Naphthalene, 2-chloro-<cr>2-Chloronaphthalene	1*	2,4	U047	D	5000 (2270)
2-Chloronaphthalene	91587	beta-Chloronaphthalene<cr>Naphthalene, 2-chloro-	1*	2,4	U047	D	5000 (2270)
2-Chlorophenol	95578	o-Chlorophenol <cr>Phenol, 2-chloro-	1*	2,4	U048	B	100 (45.4)
o-Chlorophenol	95578	Phenol, 2-chloro-<cr>2-Chlorophenol	1*	2,4	U048	B	100 (45.4)
4-Chlorophenyl phenyl ether	7005723		1*	2		D	5000 (2270)
1-(o-Chlorophenyl)thiourea	5344821	Thiourea, (2-chlorophenyl)-	1*	4	P026	B	100 (45.4)
3-Chloropropionitrile	542767	Propanenitrile, 3-chloro-	1*	4	P027	C	1000 (454)
Chlorosulfonic acid	7790945		1000	1		C	1000 (454)
4-Chloro-o-toluidine, hydrochloride	3165933	Benzenamine, 4-chloro-2--methyl-, hydrochloride	1*	4	U049	B	100 (45.4)
Chlorpyrifos	2921882		1	1		X	1 (0.454)
Chromic acetate	1066304		1000	1		C	1000 (454)

Chromic acid	11115745<cr>7738945		1000	1		A	10 (4.54)
Chromic acid H2CrO4, calcium salt	13765190	Calcium chromate	1000	1,4	U032	A	10 (4.54)
Chromic sulfate	10101538		1000	1		C	1000 (454)
Chromium	7440473		1*	2		D	5000 (2270)
CHROMIUM AND COMPOUNDS	N.A.		1*	2			**
Chromous chloride	10049055		1000	1		C	1000 (454)
Chrysene	218019	1,2-Benzphenanthrene	1*	2,4	U050	B	100 (45.4)
Cobaltous bromide	7789437		1000	1		C	1000 (454)
Cobaltous formate	544183		1000	1		C	1000 (454)
Cobaltous sulfamate	14017415		1000	1		C	1000 (454)
Coke Oven Emissions	N.A.		1*	3		X	1 (0.454)
Copper cyanide CuCN	544923	Copper cyanide	1*	4	P029	A	10 (4.54)
Copper	7440508		1*	2		D	5000 (2270)
COPPER AND COMPOUNDS	N.A.		1*	2			**
Copper cyanide	544923	Copper cyanide CuCN	1*	4	P029	A	10 (4.54)
Coumaphos	56724		10	1		A	10 (4.54)
Creosote	8001589		1*	4	U051	X	1 (0.454)
Cresol(s)	1319773	Cresylic acid<cr>Phenol, methyl-	1000	1,4	U052	C	1000 (454)
m-Cresol	108394	m-Cresylic acid					
o-Cresol	95487	o-Cresylic acid					
p-Cresol	106445	p-Cresylic acid					
Cresylic acid	1319773	Cresol(s) <cr>Phenol, methyl-	1000	1,4	U052	C	1000 (454)
m-Cresol	108394	m-Cresylic acid					
o-Cresol	95487	o-Cresylic acid					
p-Cresol	106445	p-Cresylic acid					
Crotonaldehyde	123739<cr>4170303	2-Butenal	100	1,4	U053	B	100 (45.4)

Cumene	98828	Benzene, 1-methylethyl-	1*	4	U055	D	5000 (2270)
Cupric acetate	142712		100	1		B	100 (45.4)
Cupric acetoarsenite	12002038		100	1		X	1 (0.454)
Cupric chloride	7447394		10	1		A	10 (4.54)
Cupric nitrate	3251238		100	1		B	100 (45.4)
Cupric oxalate	5893663		100	1		B	100 (45.4)
Cupric sulfate	7758987		10	1		A	10 (4.54)
Cupric sulfate, ammoniated	10380297		100	1		B	100 (45.4)
Cupric tartrate	815827		100	1		B	100 (45.4)
CYANIDES	N.A.		1*	2			**
Cyanides (soluble salts and complexes) not otherwise specified	57125		1*	4	P030	A	10 (4.54)
Cyanogen	460195	Ethanedinitrile	1*	4	P031	B	100 (45.4)
Cyanogen bromide	506683	Cyanogen bromide (CN)Br	1*	4	U246	C	1000 (454)
Cyanogen bromide (CN)Br	506683	Cyanogen bromide	1*	4	U246	C	1000 (454)
Cyanogen chloride	506774	Cyanogen chloride (CN)Cl	10	1,4	P033	A	10 (4.54)
Cyanogen chloride (CN)Cl	506774	Cyanogen chloride	10	1,4	P033	A	10 (4.54)
2,5-Cyclohexadiene-1,4-dione	106514	p-Benzoquinone	1*	4	U197	A	10 (4.54)
Cyclohexane	110827	Benzene, hexahydro-	1000	1,4	U056	C	1000 (454)
Cyclohexane, 1,2,3,4,5,6-hexachloro-, (1alpha,2alpha,3beta,4alpha,5alpha, 6beta)-	58899	gamma-BHC <cr>Hexachlorocyclohexane (gamma isomer)<cr>Lindane	1	1,2,4	U129	X	1 (0.454)
Cyclohexanone	108941		1*	4	U057	D	5000 (2270)
2-Cyclohexyl-4,6-dinitrophenol	131895	Phenol, 2-cyclohexyl-4,6-dinitro-	1*	4	P034	B	100 (45.4)
1,3-Cyclopentadiene, 1,2,3,4,5,5-hexachloro-	77474	Hexachlorocyclopentadiene	1	1,2,4	U130	A	10 (4.54)

Cyclophosphamide	50180	2H-1,3,2-Oxazaphosphorin-2-amine, <cr>N,N-bis(2-chloroethyl)tetrahydro-,2--oxide	1*	4	U058	A	10 (4.54)
2,4-D Acid	94757	Acetic acid (2,4-dichlorophenoxy)-2,4-D, salts and esters	100	1,4	U240	B	100 (45.4)
2,4-D Ester	94111 <cr>94791<cr>94804<cr>1320189<cr>1928387<cr>1928616<cr>1929733 <cr>2971382<cr>25168267 <cr>53467111		100	1		B	100 (45.4)
2,4-D, salts and esters	94757	Acetic acid (2,4-dichlorophenoxy)-2,4-D Acid	100	1,4	U240	B	100 (45.4)
Daunomycin	20830813	5,12-Naphthacenedione,8-acetyl-10-[3-amino-2,3,6-trideoxy-alpha-L-lyxo-hexo-pyranosyl)oxy]-7,8,9,10-tetrahydro-6,8,11-trihydroxy-1-methoxy-,(8S-cis)-	1*	4	U059	A	10 (4.54)
DDD	72548	Benzene, 1,1'-(2,2-dichloroethylidene)bis[4-chloro--<cr>TDE <cr>4,4' DDD	1	1,2,4	U060	X	1 (0.454)

4,4' DDD	72548	Benzene, 1,1'-(2,2-dichloroethylidene)bis[4-chloro- - <cr> DDD<cr> TDE	1	1,2,4	U060	X	1 (0.454)
DDE	72559	4,4' DDE	1*	2		X	1 (0.454)
4,4' DDE	72559	DDE	1*	2		X	1 (0.454)
DDT	50293	Benzene, 1,1'-(2,2,2-trichloroethylidene)bis[4-chloro- 4,4'DDT	1	1,2,4	U061	X	1 (0.454)
4,4'DDT	50293	Benzene, 1,1'-(2,2,2-trichloroethylidene)bis[4-chloro-<cr> DDT	1	1,2,4	U061	X	1 (0.454)
DDT AND METABOLITES	N.A.		1*	2			**
Diallate	2303164	Carbamothioic acid, bis(1-methylethyl)-, S-(2,3-dichloro-2-propenyl) ester	1*	4	U062	B	100 (45.4)
Diazinon	333415		1	1		X	1 (0.454)
Dibenz[a,h]anthracene	53703	Dibenzo[a,h]anthracene<cr> 1,2:5,6-Dibenzanthracene	1*	2,4	U063	X	1 (0.454)
1,2:5,6-Dibenzanthracene	53703	Dibenz[a,h]anthracene<cr> Dibenzo[a,h]anthracene	1*	2,4	U063	X	1 (0.454)
Dibenzo[a,h]anthracene	53703	Dibenz[a,h]anthracene<cr> 1,2:5,6-Dibenzanthracene	1*	2,4	U063	X	1 (0.454)
Dibenz[a,i]pyrene	189559	Benzo[rst]pentaphene	1*	4	U064	A	10 (4.54)
1,2-Dibromo-3-chloropropane	96128	Propane, 1,2-dibromo-3-chloro-	1*	4	U066	X	1 (0.454)

Dibutyl phthalate	84742	Di-n-butyl phthalate<cr> n-Butyl phthalate<cr> 1,2-Benzenedicarboxylic acid, dibutyl ester	100	1,2,4	U069	A	10 (4.54)
Di-n-butyl phthalate	84742	Dibutyl phthalate <cr> n-Butyl phthalate<cr> 1,2-Benzenedicarboxylic acid, dibutyl ester	100	1,2,4	U069	A	10 (4.54)
Dicamba	1918009		1000	1		C	1000 (454)
Dichlobenil	1194656		1000	1		B	100 (45.4)
Dichlone	117806		1	1		X	1 (0.454)
Dichlorobenzene	25321226		100	1		B	100 (45.4)
1,2-Dichlorobenzene	95501	Benzene, 1,2-dichloro- o-Dichlorobenzene	100	1,2,4	U070	B	100 (45.4)
1,3-Dichlorobenzene	541731	Benzene, 1,3-dichloro m-Dichlorobenzene	1*	2,4	U071	B	100 (45.4)
1,4-Dichlorobenzene	106467	Benzene, 1,4-dichloro p-Dichlorobenzene	100	1,2,4	U072	B	100 (45.4)
m-Dichlorobenzene	541731	Benzene, 1,3-dichloro 1,3-Dichlorobenzene	1*	2,4	U071	B	100 (45.4)
o-Dichlorobenzene	95501	Benzene, 1,2-dichloro 1,2-Dichlorobenzene	100	1,2,4	U070	B	100 (45.4)
p-Dichlorobenzene	106467	Benzene, 1,4-dichloro 1,4-Dichlorobenzene	100	1,2,4	U072	B	100 (45.4)
DICHLOROBENZIDINE	N.A.		1*	2			**

3,3'-Dichlorobenzidine	91941	[1,1'-Biphenyl]-4,4' diamine,3,3'dichloro-	1*	2,4	U073	X	1 (0.454)
Dichlorobromomethane	75274		1*	2		D	5000 (2270)
1,4-Dichloro-2-butene	764410	2-Butene, 1,4-dichloro-	1*	4	U074	X	1 (0.454)
Dichlorodifluoromethane	75718	Methane, dichlorodifluoro-	1*	4	U075	D	5000 (2270)
1,1-Dichloroethane	75343	Ethane, 1,1-dichloro-<cr> Ethylidene dichloride	1*	2,4	U076	C	1000 (454)
1,2-Dichloroethane	107062	Ethane, 1,2-dichloro-<cr> Ethylene dichloride	5000	1,2,4	U077	B	100 (45.4)
1,1-Dichloroethylene	75354	Ethene, 1,1-dichloro-<cr> Vinylidene chloride	5000	1,2,4	U078	B	100 (45.4)
1,2-Dichloroethylene	156605	Ethene, 1,2-dichloro- (E)	1*	2,4	U079	C	1000 (454)
Dichloroethyl ether	111444	Bis (2-chloroethyl) ether<cr> Ethane, 1,1'-oxybis[2-chloro--	1*	2,4	U025	A	10 (4.54)
Dichloroisopropyl ether	108601	Propane, 2,2'-oxybis[2-chloro--	1*	2,4	U027	C	1000 (454)
Dichloromethoxy ethane	111911	Bis(2-chloroethoxy) methane<cr> Ethane, 1,1'-[methylenebis(oxy)]bis(2-chloro-	1*	2,4	U024	C	1000 (454)
Dichloromethyl ether	542881	Methane, oxybis(chloro-	1*	4	P016	A	10 (4.54)
2,4-Dichlorophenol	120832	Phenol, 2,4-dichloro-	1*	2,4	U081	B	100 (45.4)
2,6-Dichlorophenol	87650	Phenol, 2,6-dichloro-	1*	4	U082	B	100 (45.4)
Dichlorophenylarsine	696286	Arsonous dichloride, phenyl-	1*	4	P036	X	1 (0.454)

Dichloropropane	26638197		5000	1		C	1000 (454)
1,1-Dichloropropane	78999						
1,3-Dichloropropane	142289						
1,2-Dichloropropane	78875	Propane, 1,2-dichloro-<cr> Propylene dichloride	5000	1,2,4	U083	C	1000 (454)
Dichloropropane-Dichloropropene (mixture)	8003198		5000	1		B	100 (45.4)
Dichloropropene	26952238		5000	1		B	100 (45.4)
2,3-Dichloropropene	78886						
1,3-Dichloropropene	542756	1-Propene , 1,3-dichloro-	5000	1,2,4	U084	B	100 (45.4)
2,2-Dichloropropionic acid	75990		5000	1		D	5000 (2270)
Dichlorvos	62737		10	1		A	10 (4.54)
Dicofol	115322		5000	1		A	10 (4.54)
Dieldrin	60571	2,7:3,6-Dimethanonaphth[2,3-b]oxirene, 3,4,5,6,9,9-hexachloro-1a,2,2a,3,6,6a,7,7a-octahydro-, (1aalpha,2beta,2aalpha,3beta,6beta,6aalpha,7beta,7aalpha)-	1	1,2,4	P037	X	1 (0.454)
1,2:3,4-Diepoxybutane	1464535	2,2'-Bioxirane	1*	4	U085	A	10 (4.54)
Diethylamine	109897		1000	1		B	100 (454.4)
Diethylarsine	692422	Arsine, diethyl-	1*	4	P038	X	1 (0.454)
1,4-Diethylenedioxide	123911	1,4-Dioxane	1*	4	U108	B	100 (45.4)
Diethylhexyl phthalate	117817	Bis (2-ethylhexyl)phthalate<cr> 1,2-Benzenedicarboxylic acid, [bis(2-ethylhexyl)] ester	1*	2,4	U028	B	100 (45.4)

N,N'-Diethylhydrazine	1615801	Hydrazine, 1,2-diethyl-	1*	4	U086	A	10 (4.54)
O,O-Diethyl S-methyl dithiophosphate	3288582	Phosphorodithioic acid, O,O-diethyl S-methyl ester	1*	4	U087	D	5000 (2270)
Diethyl-p-nitrophenyl phosphate	311455	Phosphoric acid, diethyl 4-nitrophenyl ester	1*	4	P041	B	100 (45.4)
Diethyl phthalate	84662	1,2-Benzenedicarboxylic acid, diethyl ester	1*	2,4	U088	C	1000 (454)
O,O-Diethyl O-pyrazinyl phosphorothioate	297972	Phosphorothioic acid, O,O-diethyl O-pyrazinyl ester	1*	4	P040	B	100 (45.4)
Diethylstilbestrol	56531	Phenol, 4,4'-(1,2-diethyl-1,2-ethenediyl)bis-, (E)	1*	4	U089	X	1 (0.454)
Dihydrosafrole	94586	1,3-Benzodioxole, 5-propyl-	1*	4	U090	A	10 (4.54)
Diisopropylfluorophosphate	55914	Phosphorofluoridic acid, bis(1-methylethyl) ester	1*	4	P043	B	100 (45.4)
1,4,5,8-Dimethanonaphthalene, 1,2,3,4,10,10-10--hexachloro-1,4,4a,5,8,8a-hexahydro-, (1alpha,4alpha,4abeta,5alpha,8alpha,	309002	Aldrin	1	1,2,4	P004	X	1 (0.454)
8abeta)-1,4,5,8-Dimethanonaphthalene, 1,2,3,4,10,10-hexachloro-1,4,4a,5,8,8a-hexahydro, (1alpha,4alpha,4abeta,5abeta,8beta,	465736	Isodrin	1*	4	P060	X	1 (0.454)

8abeta)-2,7:3,6-Dimethanonaphth[2,3-b]oxirene, 3,4,5,6,9,9-hexachloro-1a,2,2a,3,6,6a,7,7a-octahydro-, (1aalpha,2beta,2aalpha,3beta,6beta,	60571	Dieldrin	1	1,2,4	P037	X	1 (0.454)
6aalpha,7beta,7aalpha)-2,7:3,6-Dimethanonaphth[2,3-b]oxirene, 3,4,5,6,9,9-hexachloro-1a,2,2a,3,6,6a,7,7a-octa-hydro-, (1aalpha,2beta,2abeta,3alpha,6alpha,	72208	Endrin<cr > Endrin, & metabolites	1	1,2,4	P051	X	1 (0.454)
6abeta,7beta,7aalpha)-Dimethoate	60515	Phosphorodithioic acid, O,O-dimethyl S-[2(methylamino)-2-oxoethyl] ester	1*	4	P044	A	10 (4.54)
3,3'-Dimethoxybenzidine	119904	[1,1'-Biphenyl]-4,4' diamine,3,3'dimethoxy-	1*	4	U091	B	100 (45.4)
Dimethylamine	124403	Methanamine, N-methyl-	1000	1,4	U092	C	1000 (454)
p-Dimethylaminoazobenzene	60117	Benzenamine, N,N-dimethyl-4-(phenylazo-)	1*	4	U093	A	10 (4.54)
7,12-Dimethylbenz[a]anthracene	57976	Benz[a]anthracene, 7,12-dimethyl-	1*	4	U094	X	1 (0.454)
3,3'-Dimethylbenzidine	119937	[1,1'Biphenyl]-4,4'-diamine,3,3'-dimethyl-	1*	4	U095	A	10 (4.54)
alpha,alpha-Dimethylbenzylhydroperoxide	80159	Hydroperoxide, 1-methyl-1-phenylethyl-	1*	4	U096	A	10 (4.54)
Dimethylcarbamoyl chloride	79447	Carbamic chloride, dimethyl-	1*	4	U097	X	1 (0.454)
1,1-Dimethylhydrazine	57147	Hydrazine, 1,1-dimethyl-	1*	4	U098	A	10 (4.54)

1,2-Dimethylhydrazine	540738	Hydrazine, 1,2-dimethyl-	1*	4	U099	X	1 (0.454)
alpha,alpha-Dimethylphenethylamine	122098	Benzeneethanamine, alpha,alpha-dimethyl-	1*	4	P046	D	5000 (2270)
2,4-Dimethylphenol	105679	Phenol, 2,4-dimethyl-	1*	2,4	U101	B	100 (45.4)
Dimethyl phthalate	131113	1,2-Benzenedicarboxylic acid, dimethyl ester	1*	2,4	U102	D	5000 (2270)
Dimethyl sulfate	77781	Sulfuric acid, dimethyl ester	1*	4	U103	B	100 (45.4)
Dinitrobenzene (mixed)	25154545		1000	1		B	100 (45.4)
m-Dinitrobenzene	99650						
o-Dinitrobenzene	528290						
p-Dinitrobenzene	100254						
4,6-Dinitro-o-cresol and salts	534521	Phenol, 2-methyl-4,6-dinitro--	1*	2,4	P047	A	10 (4.54)
Dinitrophenol	25550587		1000	1		A	10 (4.54)
2,5-Dinitrophenol	329715						
2,6-Dinitrophenol	573568						
2,4-Dinitrophenol	51285	Phenol, 2,4-dinitro--	1000	1,2,4	P048	A	10 (4.54)
Dinitrotoluene	25321146		1000	1,2		A	10 (4.54)
3,4-Dinitrotoluene	610399						
2,4-Dinitrotoluene	121142	Benzene, 1-methyl-2,4-dinitro--	1000	1,2,4	U105	A	10 (4.54)
2,6-Dinitrotoluene	606202	Benzene, 2-methyl-1,3-dinitro--	1000	1,2,4	U106	B	100 (45.4)
Dinoseb	88857	Phenol, 2-(1-methylpropyl)-4,6-dinitro	1*	4	P020	C	1000 (454)
Di-n-octyl phthalate	117840	1,2-Benzenedicarboxylic acid, dioctyl ester	1*	2,4	U107	D	5000 (2270)

1,4-Dioxane	123911	1,4-Diethylenedioxide	1*	4	U108	B	100 (45.4)
DIPHENYLHYDRAZINE	N.A.		1*	2			**
1,2-Diphenylhydrazine	122667	Hydrazine, 1,2-diphenyl-	1*	2,4	U109	A	10 (4.54)
Diphosphoramide, octamethyl-	152169	Octamethylpyrophosphoramide	1*	4	P085	B	100 (45.4)
Diphosphoric acid, tetraethyl ester	107493	Tetraethylpyrophosphate	100	1,4	P111	A	10 (4.54)
Dipropylamine	142847	1-Propanamine, N-propyl-	1*	4	U110	D	5000 (2270)
Di-n-propylnitrosamine	621647	1-Propanamine, N-nitroso-N-propyl-	1*	2,4	U111	A	10 (4.54)
Diquat	85007 <cr>2764729		1000	1		C	1000 (454)
Disulfoton	298044	Phosphorodithioic acid, o,o-diethyl S-[2-(ethylthio)ethyl] ester	1	1,4	P039	X	1 (0.454)
Dithiobiuret	541537	Thioimidodicarbonic diamide [(H2N)<cr> C(S)]2NH	1*	4	P049	B	100 (45.4)
Diuron	330541		100	1		B	100 (45.4)
Dodecylbenzenesulfonic acid	27176870		1000	1		C	1000 (454)
Endosulfan	115297	6,9-Methano-2,4,3-benzodioxathiepin, 6,7,8,9,10,10-hexachloro-1,5,5a,6,9,9a-hexahydro-, 3-oxide	1	1,2,4	P050	X	1 (0.454)
alpha - Endosulfan	959988		1*	2		X	1 (0.454)
beta - Endosulfan	33213659		1*	2		X	1 (0.454)
ENDOSALFAN AND METABOLITES	N.A.		1*	2			**
Endosulfan sulfate	1031078		1*	2		X	1 (0.454)

Endothall	145733	7-Oxabicyclo[2.2.1]heptane-2,3-dicarboxylic acid	1*	4	P088	C	1000 (454)
Endrin	72208	Endrin, & metabolites<cr> 2,7:3,6-Dimethanonaphth[2,3-b]oxirene, 3,4,5,6,9,9 -hexachloro-1a,2,2a,3,<cr> 6,6a,7,7a--octa-hydro-, (1aalpha,<cr> 2beta,2abeta,3alpha,6alpha,<cr> 6abeta,7beta, 7aalpha)-	1	1,2,4	P051	X	1 (0.454)
Endrin aldehyde	7421934		1*	2		X	1 (0.454)
ENDRIN AND METABOLITES	N.A.		1*	2			**
Endrin, & metabolites	72208	Endrin<cr> 2,7:3,6-Dimethanonaphth[2,3-b]oxirene, <cr> 3,4,5,6,9,9-hexachloro-1a,2,2a,3,<cr> 6,6a,7,7a--octa-hydro-, (1aalpha,<cr> 2beta,2abeta,3alpha,6alpha,<cr> 6abeta,7beta, 7aalpha)-	1	1,2,4	P051	X	1 (0.454)
Epichlorohydrin	106898	Oxirane, (chloromethyl)-	1000	1,4	U041	B	100 (45.4)
Epinephrine	51434	1,2-Benzenediol,4-[1-hydroxy-2-(methylamino)ethyl]-	1*	4	P042	C	1000 (454)

Ethanal	75070	Acetaldehyde	1000	1,4	U001	C	1000 (454)
Ethanamine, N-ethyl-N-nitroso--	55185	N-Nitrosodiethylamine	1*	4	U174	X	1 (0.454)
1,2-Ethanediamine, N,N-dimethyl-N'-2-pyridinyl-N'-(2-thienylmethyl)-	91805	Methapyrilene	1*	4	U155	D	5000 (2270)
Ethane, 1,2-dibromo-	106934	Ethylene dibromide	1000	1,4	U067	X	1 (0.454)
Ethane, 1,1-dichloro-	75343	Ethylidene dichloride <cr>1,1-Dichloroethane	1*	2,4	U076	C	1000 (454)
Ethane, 1,2-dichloro-	107062	Ethylene dichloride <cr>1,2-Dichloroethane	5000	1,2,4	U077	B	100 (45.4)
Ethanedinitrile	460195	Cyanogen	1*	4	P031	B	100 (45.4)
Ethane, hexachloro-	67721	Hexachloroethane	1*	2,4	U131	B	100 (45.4)
Ethane, 1,1'-[methylenebis(oxy)]bis(2-<cr> chloro-	111911	Bis(2-chloroethoxy) methane <cr>Dichloromethoxy ethane	1*	2,4	U024	C	1000 (454)
Ethane, 1,1'-oxybis-	60297	Ethyl ether	1*	4	U117	B	100 (45.4)
Ethane, 1,1'-oxybis[2-chloro-	111444	Bis (2-chloroethyl) ether <cr>Dichloroethyl ether	1*	2,4	U025	A	10 (4.54)
Ethane, pentachloro-	76017	Pentachloroethane	1*	4	U184	A	10 (4.54)
Ethane, 1,1,1,2-tetrachloro-	630206	1,1,1,2-Tetrachloroethane	1*	4	U208	B	100 (45.4)
Ethane, 1,1,2,2-tetrachloro-	79345	1,1,2,2-Tetrachloroethane	1*	2,4	U209	B	100 (45.4)
Ethanethioamide	62555	Thioacetamide	1*	4	U218	A	10 (4.54)
Ethane, 1,1,1-trichloro-	71556	Methyl chloroform <cr>1,1,1--Trichloroethane	1*	2,4	U226	C	1000 (454)
Ethane, 1,1,2-trichloro-	79005	1,1,2-Trichloroethane	1*	2,4	U227	B	100 (45.4)
Ethanimidothioic acid, N-[[(methyl-amino)carbonyl]oxy]-, methyl ester	16752775	Methomyl	1*	4	P066	B	100 (45.4)

Ethanol, 2-ethoxy-	110805	Ethylene glycol monoethyl ether	1*	4	U359	C	1000 (454)
Ethanol, 2,2'-(nitrosoimino)bis-	1116547	N-Nitrosodiethanolamine	1*	4	U173	X	1 (0.454)
Ethanone, 1-phenyl-	98862	Acetophenone	1*	4	U004	D	5000 (2270)
Ethene, chloro-	75014	Vinyl chloride	1*	2,3,4	U043	X	1 (0.454)
Ethene, 2-chloroethoxy-	110758	2-Chloroethyl vinyl ether	1*	2,4	U042	C	1000 (454)
Ethene, 1,1-dichloro-	75354	Vinylidene chloride<cr>1,1-Dichloroethylene	5000	1,2,4	U078	B	100 (45.4)
Ethene, 1,2-dichloro- (E)	156605	1,2-Dichloroethylene	1*	2,4	U079	C	1000 (454)
Ethene, tetrachloro-	127184	Perchloroethylene<cr>Tetrachloroethene<cr>Tetrachloroethylene	1*	2,4	U210	B	100 (45.4)
Ethene, trichloro-	79016	Trichloroethene<cr>Trichloroethylene	1000	1,2,4	U228	B	100 (45.4)
Ethion	563122		10	1		A	10 (4.54)
Ethyl acetate	141786	Acetic acid, ethyl ester	1*	4	U112	D	5000 (2270)
Ethyl acrylate	140885	2-Propenoic acid, ethyl ester	1*	4	U113	C	1000 (454)
Ethylbenzene	100414		1000	1,2		C	1000 (454)
Ethyl carbamate (urethane)	51796	Carbamic acid, ethyl ester	1*	4	U238	B	100 (45.4)
Ethyl cyanide	107120	Propanenitrile	1*	4	P101	A	10 (4.54)
Ethylenebisdithiocarbamic acid, salts & esters	111546	Carbamodithioic acid, 1,2-ethanediylbis, salts & esters	1*	4	U114	D	5000 (2270)
Ethylenediamine	107153		1000	1		D	5000 (2270)
Ethylenediamine-tetraacetic acid (EDTA)	60004		5000	1		D	5000 (2270)
Ethylene dibromide	106934	Ethane, 1,2-dibromo-	1000	1,4	U067	X	1 (0.454)

Ethylene dichloride	107062	Ethane, 1,2-dichloro-<cr>1,2-Dichloroethane	5000	1,2,4	U077	B	100 (45.4)
Ethylene glycol monoethyl ether	110805	Ethanol, 2-ethoxy-	1*	4	U359	C	1000 (454)
Ethylene oxide	75218	Oxirane	1*	4	U115	A	10 (4.54)
Ethylenethiourea	96457	2-Imidazolidinethione	1*	4	U116	A	10 (4.54)
Ethylenimine	151564	Aziridine	1*	4	P054	X	1 (0.454)
Ethyl ether	60297	Ethane, 1,1'-oxybis-	1*	4	U117	B	100 (45.4)
Ethylidene dichloride	75343	Ethane, 1,1-dichloro-<cr>1,1-Dichloroethane	1*	2,4	U076	C	1000 (454)
Ethyl methacrylate	97632	2-Propenoic acid, 2-methyl-, ethyl ester	1*	4	U118	C	1000 (454)
Ethyl methanesulfonate	62500	Methanesulfonic acid, ethyl ester	1*	4	U119	X	1 (0.454)
Famphur	52857	Phosphorothioic acid, O,[4-[(dimethylamino) sulfonyl] phenyl] O,O-dimethyl ester	1*	4	P097	C	1000 (454)
Ferric ammonium citrate	1185575		1000	1		C	1000 (454)
Ferric ammonium oxalate	2944674<cr>55488874		1000	1		C	1000 (454)
Ferric chloride	7705080		1000	1		C	1000 (454)
Ferric fluoride	7783508		100	1		B	100 (45.4)
Ferric nitrate	10421484		1000	1		C	1000 (454)
Ferric sulfate	10028225		1000	1		C	1000 (454)
Ferrous ammonium sulfate	10045893		1000	1		C	1000 (454)
Ferrous chloride	7758943		100	1		B	100 (45.4)
Ferrous sulfate	7720787<cr>7782630		1000	1		C	1000 (454)
Fluoranthene	206440	Benzo[j,k]fluorene	1*	2,4	U120	B	100 (45.4)

Fluorene	86737		1*	2		D	5000 (2270)
Fluorine	7782414		1*	4	P056	A	10 (4.54)
Fluoroacetamide	640197	Acetamide, 2-fluoro-	1*	4	P057	B	100 (45.4)
Fluoroacetic acid, sodium salt	62748	Acetic acid, fluoro-, sodium salt	1*	4	P058	A	10 (4.54)
Formaldehyde	50000		1000	1,4	U122	B	100 (45.4)
Formic acid	64186		5000	1,4	U123	D	5000 (2270)
Fulminic acid, mercury(2+)salt	628864	Mercury fulminate	1*	4	P065	A	10 (4.54)
Fumaric acid	110178		5000	1		D	5000 (2270)
Furan	110009	Furfuran	1*	4	U124	B	100 (45.4)
Furan, tetrahydro-	109999	Tetrahydrofuran	1*	4	U213	C	1000 (454)
2-Furancarboxaldehyde	98011	Furfural	1000	1,4	U125	D	5000 (2270)
2,5-Furandione	108316	Maleic anhydride	5000	1,4	U147	D	5000 (2270)
Furfural	98011	2-Furancarboxaldehyde	1000	1,4	U125	D	5000 (2270)
Furfuran	110009	Furan	1*	4	U124	B	100 (45.4)
Glucopyranose, 2-deoxy-2-(3-methyl-3-nitrosoureido)-	18883664	D-Glucose, 2-deoxy-2--[[(methylnitrosoamino)-carbonyl]amino] Streptozotocin	1*	4	U206	X	1 (0.454)
D-Glucose, 2-deoxy-2-[[(methylnitrosoamino)-carbonyl]amino]-	18883664	Glucopyranose, 2-deoxy-2--(3-methyl--3-nitrosoureido)-<cr>Streptozotocin	1*	4	U206	X	1 (0.454)
Glycidylaldehyde	765344	Oxiranecarboxyaldehyde	1*	4	U126	A	10 (4.54)
Guanidine, N-methyl-N'-nitro-N-nitroso-	70257	MNNG	1*	4	U163	A	10 (4.54)
Guthion	86500		1	1		X	1 (0.454)
HALOETHERS	N.A.		1*	2			**
HALOMETHANES	N.A.		1*	2			**

Heptachlor	76448	4,7-Methano-1H-indene, 1,4,5,6,7,8,8-heptachloro-3a,4,7,7a-tetrahydro-	1	1,2,4	P059	X	1 (0.454)
HEPTACHLOR AND METABOLITES	N.A.		1*	2			**
Heptachlor epoxide	1024573		1*	2		X	1 (0.454)
Hexachlorobenzene	118741	Benzene, hexachloro-	1*	2,4	U127	A	10 (4.54)
Hexachlorobutadiene	87683	1,3-Butadiene, 1,1,2,3,4,4-hexachloro-	1*	2,4	U128	X	1 (0.454)
HEXACHLOROCYCLOHEXANE (all isomers)	608731		1*	2			**
Hexachlorocyclohexane (gamma isomer)	58899	Cyclohexane, 1,2,3,4,5,6-hexachloro-, (1alpha,2alpha,3beta,4alpha,5alpha,6beta)-gamma-BHC <cr>Lindane	1	1,2,4	U129	X	1 (0.454)
Hexachlorocyclopentadiene	77474	1,3-Cyclopentadiene,1,2,3,4,5,5-hexachloro-	1	1,2,4	U130	A	10 (4.54)
Hexachloroethane	67721	Ethane, hexachloro-	1*	2,4	U131	B	100 (45.4)
Hexachlorophene	70304	Phenol, 2,2'-methylenebis[3,4,6-trichloro-	1*	4	U132	B	100 (45.4)
Hexachloropropene	1888717	1-Propene, 1,1,2,3,3,3-hexachloro-	1*	4	U243	C	1000 (454)
Hexaethyl tetraphosphate	757584	Tetraphosphoric acid, hexaethyl ester	1*	4	P062	B	100 (45.4)
Hydrazine	302012		1*	4	U133	X	1 (0.454)
Hydrazine, 1,2-diethyl-	1615801	N,N'-Diethylhydrazine	1*	4	U086	A	10 (4.54)

Hydrazine, 1,1-dimethyl-	57147	1,1-Dimethylhydrazine	1*	4	U098	A	10 (4.54)
Hydrazine, 1,2-dimethyl-	540738	1,2-Dimethylhydrazine	1*	4	U099	X	1 (0.454)
Hydrazine, 1,2-diphenyl-	122667	1,2-Diphenylhydrazine	1*	2,4	U109	A	10 (4.54)
Hydrazine, methyl-	60344	Methyl hydrazine	1*	4	P068	A	10 (4.54)
Hydrazinecarbothioamide	79196	Thiosemicarbazide	1*	4	P116	B	100 (45.4)
Hydrochloric acid	7647010	Hydrogen chloride	5000	1		D	5000 (2270)
Hydrocyanic acid	74908	Hydrogen cyanide	10	1,4	P063	A	10 (4.54)
Hydrofluoric acid	7664393	Hydrogen fluoride	5000	1,4	U134	B	100 (45.4)
Hydrogen chloride	7647010	Hydrochloric acid	5000	1		D	5000 (2270)
Hydrogen cyanide	74908	Hydrocyanic acid	10	1,4	P063	A	10 (4.54)
Hydrogen fluoride	7664393	Hydrofluoric acid	5000	1,4	U134	B	100 (45.4)
Hydrogen sulfide	7783064	Hydrogen sulfide H2S	100	1,4	U135	B	100 (45.4)
Hydrogen sulfide H2S	7783064	Hydrogen sulfide	100	1,4	U135	B	100 (45.4)
Hydroperoxide, 1-methyl-1-phenylethyl-	80159	alpha,alpha-Dimethylbenzylhydroperoxide	1*	4	U096	A	10 (4.54)
2-Imidazolidinethione	96457	Ethylenethiourea	1*	4	U116	A	10 (4.54)
Indeno(1,2,3-cd) pyrene	193395	1,10-(1,2-Phenylene)pyrene	1*	2,4	U137	B	100 (45.4)
1,3-Isobenzofurandione	85449	Phthalic anhydride	1*	4	U190	D	5000 (2270)
Isobutyl alcohol	78831	1-Propanol, 2-methyl-	1*	4	U140	D	5000 (2270)
Isodrin	465736	1,4,5,8-Dimethanonaphthalene, 1,2,3,4,10,10-hexachloro-1,4,4a,5,8,8a-hexahydro, (1alpha,4alpha,4abeta,5beta,8beta,8abeta)-	1*	4	P060	X	1 (0.454)
Isophorone	78591		1*	2		D	5000 (2270)

Isoprene	78795		1000	1		B	100 (45.4)
Isopropanolamine dodecylbenzenesulfonate	42504461		1000	1		C	1000 (454)
Isosafrole	120581	1,3-Benzodioxole, 5-)1-propenyl)-	1*	4	U141	B	100 (45.4)
3(2H)-Isoxazolone, 5-(aminomethyl)-	2763964	Muscimol <cr>5-(Aminomethyl)-3-isoxazolol	1*	4	P007	C	1000 (454)
Kepone	143500	1,3,4-Metheno-2H-cyclobutal[cd]pentalen-2-one, 1,1a,3,3a,4,5,5,5a,5b,6-decachloroctahydro-	1	1,4	U142	X	1 (0.454)
Lasiocarpine	303344	2-Butenoic acid, 2-methyl-, 7[[2,3-dihydroxy-2-(1-methoxyethyl)-3-methyl-1-oxobutoxy]methyl]-2,3,5,7a-tetrahydro-1H-pyrrolizin-1-yl ester, [1S-[1alpha(Z), 7(2S*,3R*),7aalpha]]-	1*	4	U143	A	10 (4.54)
Lead	7439921		1*	2		A	10 (4.54)
Lead acetate	301042	Acetic acid, lead(2+) salt	5000	1,4	U144	A	10 (4.54)
LEAD AND COMPOUNDS	N.A.		1*	2			**
Lead arsenate	7784409 <cr>7645252 <cr>10102484		5000	1		X	1 (0.454)
Lead, bis(acetatato-O)tetrahydroxytri-	1335326	Lead subacetate	1*	4	U146	A	10 (4.54)
Lead chloride	7758954		5000	1		A	10 (4.54)
Lead fluoborate	13814965		5000	1		A	10 (4.54)
Lead fluoride	7783462		1000	1		A	10 (4.54)
Leod iodide	10101630		5000	1		A	10 (4.54)

Lead nitrate	10099748		5000	1		A	10 (4.54)
Lead phosphate	7446277	Phosphoric acid, lead(2+) salt (2:3)	1*	4	U145	A	10 (4.54)
Lead stearate	1072351<cr>7428480<cr>52652592<cr>56189094		5000	1		A	10 (4.54)
Lead subacetate	1335326	Lead, bis(acetato-O)tetrahydroxytri-	1*	4	U146	A	10 (4.54)
Lead sulfate	7446142<cr>15739807		5000	1		A	10 (4.54)
Lead sulfide	1314870		5000	1		A	10 (4.54)
Lead thiocyanate	592870		5000	1		A	10 (4.54)
Lindane	58899	Cyclohexane, 1,2,3,4,5,6-hexachloro-, (1alpha,2alpha,3beta,4alpha,5alpha,6beta)-gamma-BHC<cr>Hexachlorocyclohexane (gamma isomer)	1	1,2,4	U129	X	1 (0.454)
Lithium chromate	14307358		1000	1		A	10 (4.54)
Malathion	121755		10	1		B	100 (45.4)
Maleic acid	110167		5000	1		D	5000 (2270)
Maleic anhydride	108316	2,5-Furandione	5000	1,4	U147	D	5000 (2270)
Maleic hydrazide	123331	3,6-Pyridazinedione, 1,2-dihydro-	1*	4	U148	D	5000 (2270)
Malononitrile	109773	Propanedinitrile	1*	4	U149	C	1000 (454)
Melphalan	148823	L-Phenylalanine, 4-[bis(2-chloroethyl)aminol]	1*	4	U150	X	1 (0.454)
Mercaptodimethur	2032657		100	1		A	10 (4.54)
Mercuric cyanide	592041		1	1		X	1 (0.454)
Mercuric nitrate	10045940		10	1		A	10 (4.54)
Mercuric sulfate	7783359		10	1		A	10 (4.54)

Mercuric thiocyanate	592858		10	1		A	10 (4.54)
Mercurous nitrate	10415755 <cr>7782867		10	1		A	10 (4.54)
Mercury	7439976		1*	2,3,4	U151	X	1 (0.454)
MERCURY AND COMPOUNDS	N.A.		1*	2			**
Mercury, (acetate-O)phenyl--	62384	Phenylmercury acetate	1*	4	P092	B	100 (45.4)
Mercury fulminate	628864	Fulminic acid, mercury(2+)salt	1*	4	P065	A	10 (4.54)
Methacrylonitrile	126987	2-Propenenitrile, 2-methyl-	1*	4	U152	C	1000 (454)
Methanamine, N-methyl-	124403	Dimethylamine	1000	1,4	U092	C	1000 (454)
Methanamine, N-methyl-N-nitroso-	62759	N-Nitrosodimethylamine	1*	2,4	P082	A	10 (4.54)
Methane, bromo-	74839	Methyl bromide	1*	2,4	U029	C	1000 (454)
Methane, chloro-	74873	Methyl chloride	1*	2,4	U045	B	100 (45.4)
Methane, chloromethoxy-	107302	Chloromethyl methyl ether	1*	4	U046	A	10 (4.54)
Methane, dibromo-	74953	Methylene bromide	1*	4	U068	C	1000 (454)
Methane, dichloro-	75092	Methylene chloride	1*	2,4	U080	C	1000 (454)
Methane, dichlorodifluoro-	75718	Dichlorodifluoromethane	1*	4	U075	D	5000 (2270)
Methane, iodo-	74884	Methyl iodide	1*	4	U138	B	100 (45.4)
Methane, isocyanato-	624839	Methyl isocyanate	1*	3,4	P064	A	10 (4.54)
Methane, oxybis(chloro-	542881	Dichloromethyl ether	1*	4	P016	A	10 (4.54)
Methanesulfenyl chloride, trichloro-	594423	Trichloromethanesulfenyl chloride	1*	4	P118	B	100 (45.4)
Methanesulfonic acid, ethyl ester	62500	Ethyl methanesulfonate	1*	4	U119	X	1 (0.454)
Methane, tetrachloro-	56235	Carbon tetrachloride	5000	1,2,4	U211	A	10 (4.54)
Methane, tetranitro-	509148	Tetranitromethane	1*	4	P112	A	10 (4.54)
Methane, tribromo-	75252	Bromoform	1*	2,4	U225	B	100 (45.4)
Methane, trichloro-	67663	Chloroform	5000	1,2,4	U044	A	10 (4.54)
Methane, trichlorofluoro-	75694	Trichloromonofluoromethane	1*	4	U121	D	5000 (2270)

Methanethiol	74931	Methylmercaptan <cr>Thiomethanol	100	1,4	U153	B	100 (45.4)
6,9-Methano-2,4,3-benzodioxathiepin, 6,7,8,9,10,10-hexachloro-1,5,5a,6,9,9a-hexahydro-, 3-oxide	115297	Endosulfan	1	1,2,4	P050	X	1 (0.454)
1,3,4-Metheno-2H-cyclobutal[cd]pentalen-2-one, 1,1a,3,3a,4,5,5,5a,5b,6-decachloroctahydro-	143500	Kepone	1	1,4	U142	X	1 (0.454)
4,7-Methano-1H-indene, 1,4,5,6,7,8,8-heptachloro-3a,4,7,7a-tetrahydro-	76448	Heptachlor	1	1,2,4	P059	X	1 (0.454)
4,7-Methano-1H-indene, 1,2,4,5,6,7,8,8-octachloro-2,3,3a,4,7,7a-hexahydro-	57749	Chlordane <cr>Chlordane, alpha & gamma isomers <cr>Chlordane, technical	1	1,2,4	U036	X	1 (0.454)
Methanol	67561	Methyl alcohol	1*	4	U154	D	5000 (2270)
Methapyrilene	91805	1,2-Ethanediamine, N,N-dimethyl-N'-2-pyridinyl-N'-(2-thienylmethyl)-	1*	4	U155	D	5000 (2270)
Methomyl	16752775	Ethanimidothioic acid, N-[[(methyl-amino)carbonyl]oxy]--, methyl ester	1*	4	P066	B	100 (45.4)
Methoxychlor	72435	Benzene, 1,1'-(2,2,2-trichloroethylidene) bis[4-methoxy-	1	1,4	U247	X	1 (0.454)
Methyl alcohol	67561	Methanol	1*	4	U154	D	5000 (2270)
Methyl bromide	74839	Methane, bromo-	1*	2,4	U029	C	1000 (454)
1-Methylbutadiene	504609	1,3-Pentadiene	1*	4	U186	B	100 (45.4)

Methyl chloride	74873	Methane, chloro-	1*	2,4	U045	B	100 (45.4)
Methyl chlorocarbonate	79221	Carbonochloridic acid, methyl ester <cr>Methyl chloroformate	1*	4	U156	C	1000 (454)
Methyl chloroform	71556	Ethane, 1,1,1-trichloro- <cr>1,1,1--Trichloroethane	1*	2,4	U226	C	1000 (454)
Methyl chloroformate	79221	Carbonochloridic acid, methyl ester <cr>Methyl chlorocarbonate	1*	4	U156	C	1000 (454)
3-Methylcholanthrene	56495	Benz[j]aceanthrylene, 1,2-dihydro-3-methyl-	1*	4	U157	A	10 (4.54)
4,4'-Methylenebis (2-chloroaniline)	101144	Benzenamine, 4,4'-methylenebis(2--chloro-	1*	4	U158	A	10 (4.54)
Methylene bromide	74953	Methane, dibromo-	1*	4	U068	C	1000 (454)
Methylene chloride	75092	Methane, dichloro-	1*	2,4	U080	C	1000 (454)
Methyl ethyl ketone (MEK)	78933	2-Butanone	1*	4	U159	D	5000 (2270)
Methyl ethyl ketone peroxide	1338234	2-Butanone peroxide	1*	4	U160	A	10 (4.54)
Methyl hydrazine	60344	Hydrazine, methyl-	1*	4	P068	A	10 (4.54)
Methyl iodide	74884	Methane, iodo-	1*	4	U138	B	100 (45.4)
Methyl isobutyl ketone	108101	4-Methyl-2-pentanone	1*	4	U161	D	5000 (2270)
Methyl isocyanate	624839	Methane, isocyanato-	1*	3,4	P064	A	10 (4.54)
2-Methyllactonitrile	75865	Acetone cyanohydrin <cr>Propanenitrile, 2-hydroxy--2-methyl-	10	1,4	P069	A	10 (4.54)

Methylmercaptan	74931	Methanethiol <cr>Thiomethanol	100	1,4	U153	B	100 (45.4)
Methyl methacrylate	80626	2-Propenoic acid, 2-methyl-, methyl ester	5000	1,4	U162	C	1000 (454)
Methyl parathion	298000	Phosphorothioic acid, O,O-dimethyl O-(4-nitrophenyl) ester	100	1,4	P071	B	100 (45.4)
4-Methyl-2-pentanone	108101	Methyl isobutyl ketone	1*	4	U161	D	5000 (2270)
Methylthiouracil	56042	4(1H)-Pyrimidinone, 2,3-dihydro-6-methyl-2-thioxo-	1*	4	U164	A	10 (4.54)
Mevinphos	7786347		1	1		A	10 (4.54)
Mexacarbate	315184		1000	1		C	1000 (454)
Mitomycin C	50077	Azirino[2',3':3,4]pyrrolo[1,2-a]indole-4,7-dione,6--amino-8-[[(aminocarbonyl)oxy] methyl]-1, 1a,2,8,8a,8b-hexahydro-8a-methoxy-5-methyl-, [1aS-(1aalpha, 8beta, 8aalpha, 8balpha)]-	1*	4	U010	A	10 (4.54)
MNNG	70257	Guanidine, N-methyl-N'-nitro-N--nitroso-	1*	4	U163	A	10 (4.54)
Monoethylamine	75047		1000	1		B	100 (45.4)
Monomethylamine	74895		1000	1		B	100 (45.4)
Multi Source Leachate			1*	4	F039	X	1 (0.454)

Muscimol	2763964	3(2H)-Isoxazolone, 5-(aminomethyl)- 5-(Aminomethyl)-3-isoxazolol	1*	4	P007	C	1000 (454)
Naled	300765		10	1		A	10 (4.54)
5,12-Naphthacenedione, 8-acetyl-10-[3-amino-2,3,6-trideoxy-alpha-L-lyxo-hexopyranosyl)oxy]-7,8,9,10-tetrahydro-6,8,11-trihydroxy-1-methoxy-, (8S-cis)-	20830813	Daunomycin	1*	4	U059	A	10 (4.54)
1-Naphthalenamine	134327	alpha-Naphthylamine	1*	4	U167	B	100 (45.4)
2-Naphthalenamine	91598	beta-Naphthylamine	1*	4	U168	A	10 (4.54)
Naphthalenamine, N,N'-bis(2-chloroethyl)-	494031	Chlornaphazine	1*	4	U026	B	100 (45.4)
Naphthalene	91203		5000	1,2,4	U165	B	100 (45.4)
Naphthalene, 2-chloro-	91587	beta-Chloronaphthalene 2-Chloronaphthalene	1*	2,4	U047	D	5000 (2270)
1,4-Naphthalenedione	130154	1,4-Naphthoquinone	1*	4	U166	D	5000 (2270)
2,7-Naphthalenedisulfonic acid, 3,3'-[(3,3'-dimethyl-(1,1'-biphenyl)-4,4'-diyl)-bis(azo)]bis(5-amino-4-hydroxy)-tetrasodium salt	72571	Trypan blue	1*	4	U236	A	10 (4.54)
Naphthenic acid	1338245		100	1		B	100 (45.4)
1,4-Naphthoquinone	130154	1,4-Naphthalenedione	1*	4	U166	D	5000 (2270)
alpha-Naphthylamine	134327	1-Naphthalenamine	1*	4	U167	B	100 (45.4)
beta-Naphthylamine	91598	2-Naphthalenamine	1*	4	U168	A	10 (4.54)
alpha-Naphthylthiourea	86884	Thiourea, 1-naphthalenyl-	1*	4	P072	B	100 (45.4)
Nickel	7440020		1*	2		B	100 (45.4)
Nickel ammonium sulfate	15699180		5000	1		B	100 (45.4)

NICKEL AND COMPOUNDS	N.A.		1*	2			**
Nickel carbonyl	13463393	Nickel carbonyl Ni(CO)4, (T-4)-	1*	4	P073	A	10 (4.54)
Nickel carbonyl Ni(CO)4, (T-4)-	13463393	Nickel carbonyl	1*	4	P073	A	10 (4.54)
Nickel chloride	7718549 <cr>37211055		5000	1		B	100 (45.4)
Nickel cyanide	557197	Nickel cyanide Ni(CN)2	1*	4	P074	A	10 (4.54)
Nickel cyanide Ni(CN)2	557197	Nickel cyanide	1*	4	P074	A	10 (4.54)
Nickel hydroxide	12054487		1000	1		A	10 (4.54)
Nickel nitrate	14216752		5000	1		B	100 (45.4)
Nickel sulfate	7786814		5000	1		B	100 (45.4)
Nicotine, & salts	54115	Pyridine, 3-(1-methyl-2-pyrrolidinyl)-, (S)-	1*	4	P075	B	100 (45.4)
Nitric acid	7697372		1000	1		C	1000 (454)
Nitric acid, thallium (1+) salt	10102451	Thallium (I) nitrate	1*	4	U217	B	100 (45.4)
Nitric oxide	10102439	Nitrogen oxide NO	1*	4	P076	A	10 (4.54)
p-Nitroaniline	100016	Benzenamine, 4-nitro-	1*	4	P077	D	5000 (2270)
Nitrobenzene	98953	Benzene, nitro-	1000	1,2,4	U169	C	1000 (454)
Nitrogen dioxide	10102440 <cr>10544726	Nitrogen oxide NO2	1000	1,4	P078	A	10 (4.54)
Nitrogen oxide NO	10102439	Nitric oxide	1*	4	P076	A	10 (4.54)
Nitrogen oxide NO2	10102440 <cr>10544726	Nitrogen dioxide	1000	1,4	P078	A	10 (4.54)
Nitroglycerine	55630	1,2,3-Propanetriol, trinitrate-	1*	4	P081	A	10 (4.54)
Nitrophenol (mixed)	25154556		1000	1		B	100 (45.4)
m-Nitrophenol	554847					B	100 (45.4)
o-Nitrophenol	88755	2-Nitrophenol					
p-Nitrophenol	100027	Phenol, 4-nitro- <cr> 4-Nitrophenol					

o-Nitrophenol	88755	2-Nitrophenol	1000	1,2		B	100 (45.4)
p-Nitrophenol	100027	Phenol, 4-nitro- 4-Nitrophenol	1000	1,2,4	U170	B	100 (45.4)
2-Nitrophenol	88755	o-Nitrophenol	1000	1,2		B	100 (45.4)
4-Nitrophenol	100027	p-Nitrophenol Phenol, 4-nitro-	1000	1,2,4	U170	B	100 (45.4)
NITROPHENOLS	N.A.		1*	2			**
2-Nitropropane	79469	Propane, 2-nitro-	1*	4	U171	A	10 (4.54)
NITROSAMINES	N.A.		1*	2			**
N-Nitrosodi-n-butylamine	924163	1-Butanamine, N-butyl-N--nitroso-	1*	4	U172	A	10 (4.54)
N-Nitrosodiethanolamine	1116547	Ethanol, 2,2'-(nitrosoimino)bis-	1*	4	U173	X	1 (0.454)
N-Nitrosodiethylamine	55185	Ethanamine, N-ethyl-N--nitroso-	1*	4	U174	X	1 (0.454)
N-Nitrosodimethylamine	62759	Methanamine, N-methyl-N-nitroso-	1*	2,4	P082	A	10 (4.54)
N-Nitrosodiphenylamine	86306		1*	2		B	100 (45.4)
N-Nitroso-N-ethylurea	759739	Urea, N-ethyl-N--nitroso-	1*	4	U176	X	1 (0.454)
N-Nitroso-N-methylurea	684935	Urea, N-methyl-N-nitroso	1*	4	U177	X	1 (0.454)
N-Nitroso-N-methylurethane	615532	Carbamic acid, methylnitroso-, ethyl ester	1*	4	U178	X	1 (0.454)
N-Nitrosomethylvinylamine	4549400	Vinylamine, N-methyl-N-nitroso-	1*	4	P084	A	10 (4.54)
N-Nitrosopiperidine	100754	Piperidine, 1-nitroso-	1*	4	U179	A	10 (4.54)
N-Nitrosopyrrolidine	930552	Pyrrolidine, 1-nitroso-	1*	4	U180	X	1 (0.454)
Nitrotoluene	1321126		1000	1		C	1000 (454)
m-Nitrotoluene	99081						
o-Nitrotoluene	88722						
p-Nitrotoluene	99990						

5-Nitro-o-toluidine	99558	Benzenamine, 2-methyl-5-nitro-	1*	4	U181	B	100 (45.4)
Octamethylpyrophosphoramide	152169	Diphosphoramide, octamethyl--	1*	4	P085	B	100 (45.4)
Osmium oxide OsO4 (T-4)-	20816120	Osmium tetroxide	1*	4	P087	C	1000 (454)
Osmium tetroxide	20816120	Osmium oxide OsO4 (T-4)-	1*	4	P087	C	1000 (454)
7-Oxabicyclo[2.2.1]heptane-2,3-dicarboxylic acid	145733	Endothall	1*	4	P088	C	1000 (454)
1,2-Oxathiolane, 2,2-dioxide	1120714	1,3-Propane sultone	1*	4	U193	A	10 (4.54)
2H-1,3,2-Oxazaphosphorin-2-amine, N,N-bis(2-chloroethyl)tetrahydro-, 2-oxide	50180	Cyclophosphamide	1*	4	U058	A	10 (4.54)
Oxirane	75218	Ethylene oxide	1*	4	U115	A	10 (4.54)
Oxiranecarboxyaldehyde	765344	Glycidylaldehyde	1*	4	U126	A	10 (4.54)
Oxirane, (chloromethyl)-	106898	Epichlorohydrin	1000	1,4	U041	B	100 (45.4)
Paraformaldehyde	30525894		1000	1		C	1000 (454)
Paraldehyde	123637	1,3,5-Trioxane, 2,4,6-trimethyl-	1*	4	U182	C	1000 (454)
Parathion	56382	Phosphorothioic acid, O,O-diethyl O-(4-nitrophenyl) ester	1	1,4	P089	A	10 (4.54)
Pentachlorobenzene	608935	Benzene, pentachloro-	1*	4	U183	A	10 (4.54)
Pentachloroethane	76017	Ethane, pentachloro-	1*	4	U184	A	10 (4.54)
Pentachloronitrobenzene (PCNB)	82688	Benzene, pentachloronitro-	1*	4	U185	B	100 (45.4)
Pentachlorophenol	87865	Phenol, pentachloro-	10	1,2,4	U242	A	10 (4.54)
1,3-Pentadiene	504609	1-Methylbutadiene	1*	4	U186	B	100 (45.4)

Perchloroethylene	127184	Ethene, tetrachloro- - Tetrachlor o- ethene Tetrachlor oethylene	1*	2,4	U210	B	100 (45.4)
Phenacetin	62442	Acetamide , N-(4-etho xyphenyl)-	1*	4	U187	B	100 (45.4)
Phenanthrene	85018		1*	2		D	5000 (2270)
Phenol	108952	Benzene, hydroxy-	1000	1,2,4	U188	C	1000(454)
Phenol, 2-chloro-	95578	o-Chlorop henol 2-Chlorop henol	1*	2,4	U048	B	100 (45.4)
Phenol, 4-chloro-3-methyl- -.	59507	p-Chloro- m-cresol<cr>4-Chlo ro-m-cres ol	1*	2,4	U039	D	5000 (2270)
Phenol, 2-cyclohexyl-4,6- dinitro-	131895	2-Cyclohe xyl-4,6-di nitrophen ol	1*	4	P034	B	100 (45.4)
Phenol, 2,4-dichloro-	120832	2,4-Dichlo rophenol	1*	2,4	U081	B	100 (45.4)
Phenol, 2,6-dichloro-	87650	2,6-Dichlo rophenol	1*	4	U082	B	100 (45.4)
Phenol, 4,4'-(1,2-diethyl- 1,2-ethenediyl)bis- -, (E)	56531	Diethylstil bestrol	1*	4	U089	X	1 (0.454)
Phenol, 2,4-dimethyl-	105679	2,4-Dimet hylphenol	1*	2,4	U101	B	100(45.4)
Phenol, 2,4-dinitro-	51285	2,4-Dinitr ophenol	1000	1,2,4	P048	A	10 (4.54)
Phenol, methyl-	1319773	Cresol(s) Cresylic acid	1000	1,4	U052	C	1000 (454)
m-Cresol	108394	m-Cresylic acid					
o-Cresol	95487	o-Cresylic acid					
p-Cresol	106445	p-Cresylic acid					
Phenol, 2-methyl-4,6-dinit ro-	534521	4,6-Dinitr o-o-cresol and salts	1*	2,4	P047	A	10 (4.54)
Phenol, 2,2'-methylenebis [3,4,6-trichloro-	70304	Hexachlor ophene	1*	4	U132	B	100 (45.4)
Phenol, 2-(1-methylpropyl)-4,6-dinitro	88857	Dinoseb	1*	4	P020	C	1000 (454)
Phenol, 4-nitro-	100027	p-Nitroph enol<cr>4- -Nitrophe nol	1000	1,2,4	U170	B	100 (45.4)

Phenol, pentachloro-	87865	Pentachlorophenol	10	1,2,4	U242	A	10 (4.54)
Phenol, 2,3,4,6-tetrachloro-	58902	2,3,4,6-Tetrachlorophenol	1*	4	U212	A	10 (4.54)
Phenol, 2,4,5-trichloro-	95954	2,4,5-Trichlorophenol	10	1,4	U230	A	10 (4.54)
Phenol, 2,4,6-trichloro-	88062	2,4,6-Trichlorophenol	10	1,2,4	U231	A	10 (4.54)
Phenol, 2,4,6-trinitro-, ammonium salt	131748	Ammonium picrate	1*	4	P009	A	10 (4.54)
L-Phenylalanine, 4-[bis(2-chloroethyl) aminol]	148823	Melphalan	1*	4	U150	X	1 (0.454)
1,10-(1,2-Phenylene)pyrene	193395	Indeno(1,2,3-cd)pyrene	1*	2,4	U137	B	100 (45.4)
Phenylmercury acetate	62384	Mercury, (acetato-O)phenyl-	1*	4	P092	B	100 (45.4)
Phenylthiourea	103855	Thiourea, phenyl-	1*	4	P093	B	100 (45.4)
Phorate	298022	Phosphorodithioic acid, O,O-diethyl S-(ethylthio), methyl ester	1*	4	P094	A	10 (4.54)
Phosgene	75445	Carbonic dichloride	5000	1,4	P095	A	10 (4.54)
Phosphine	7803512		1*	4	P096	B	100 (45.4)
Phosphoric acid	7664382		5000	1		D	5000 (2270)
Phosphoric acid, diethyl 4-nitrophenyl ester	311455	Diethyl-p-nitrophenyl phosphate	1*	4	P041	B	100 (45.4)
Phosphoric acid, lead(2+) salt (2:3)	7446277	Lead phosphate	1*	4	U145	A	10 (4.54)
Phosphorodithioic acid, O,O-diethyl S-[2-(ethylthio)ethyl]ester	298044	Disulfoton	1	1,4	P039	X	1 (0.454)
Phosphorodithioic acid, O,O-diethyl S-(ethylthio), methyl ester	298022	Phorate	1*	4	P094	A	10 (4.54)
Phosphorodithioic acid, O,O-diethyl S-methyl ester	3288582	O,O-Diethyl S-methyl dithiophosphate	1*	4	U087	D	5000 (2270)
Phosphorodithioic acid, O,O-dimethyl S-[2(methylamino)-2-oxoethyl] ester	60515	Dimethoate	1*	4	P044	A	10 (4.54)

Phosphorofluoridic acid, bis(1-methylethyl) ester	55914	Diisopropylfluorophosphate	1*	4	P043	B	100 (45.4)
Phosphorothioic acid, O,O-diethyl O-(4-nitrophenyl) ester	56382	Parathion	1	1,4	P089	A	10 (4.54)
Phosphorothioic acid, O,[4-[(dimethylamino) sulfonyl]phenyl]O,O-dimethyl ester	52857	Famphur	1*	4	P097	C	1000 (454)
Phosphorothioic acid, O,O-dimethyl O-(4-nitrophenyl) ester	298000	Methyl parathion	100	1,4	P071	B	100 (45.4)
Phosphorothioic acid, O,O-diethyl O-pyrazinyl ester	297972	O,O-Diethyl O-pyrazinyl phosphorothioate	1*	4	P040	B	100 (45.4)
Phosphorus	7723140		1	1		X	1 (0.454)
Phosphrous oxycloride	10025873		5000	1		C	1000 (454)
Phosphorus pentasulfide	1314803	Phosphorus sulfide Sulfur phosphide	100	1,4	U189	B	100 (45.4)
Phosphorus sulfide	1314803	Phosphorus pentasulfide Sulfur phosphide	100	1,4	U189	B	100 (45.4)
Phosphorus trichloride	7719122		5000	1		C	1000 (454)
PHTHALATE ESTERS	N.A.		1*	2			**
Phthalic anhydride	85449	1,3-Isobenzofurandione	1*	4	U190	D	5000 (2270)
2-Picoline	109068	Pyridine, 2-methyl-	1*	4	U191	D	5000 (2270)
Piperidine, 1-nitroso-	100754	N-Nitrosopiperidine	1*	4	U179	A	10 (4.54)
Plumbane, tetraethyl-	78002	Tetraethyl lead	100	1,4	P110	A	10 (4.54)
POLYCHLORINATED BIPHENYLS (PCBs)	1336363		10	1,2		X	1 (0.454)
Aroclor 1016	12674112	POLYCHLORINATED BIPHENYLS (PCBs)					
Aroclor 1221	11104282	POLYCHLORINATED BIPHENYLS (PCBs)					

Aroclor 1232	1114116 5	POLYCHL ORINATE D BIPHENYL S (PCBs)					
Aroclor 1242	5346921 9	POLYCHL ORINATE D BIPHENYL S (PCBs)					
Aroclor 1248	1267229 6	POLYCHL ORINATE D BIPHENYL S (PCBs)					
Aroclor 1254	1109769 1	POLYCHL ORINATE D BIPHENYL S (PCBs)					
Aroclor 1260	1109682 5	POLYCHL ORINATE D BIPHENYL S (PCBs)					
POLYNUCLEAR AROMATIC HYDROCARBON S	N.A.		1*	2			**
Potassium arsenate	7784410		1000	1		X	1 (0.454)
Potassium arsenite	1012450 2		1000	1		X	1 (0.454)
Potassium bichromate	7778509		1000	1		A	10 (4.54)
Potassium chromate	7789006		1000	1		A	10 (4.54)
Potassium cyanide	151508	Potassium cyanide K (CN)	10	1,4	P098	A	10 (4.54)
Potassium cyanide K(CN)	151508	Potassium cyanide	10	1,4	P098	A	10 (4.54)
Potassium hydroxide	1310583		1000	1		C	1000 (454)
Potassium permanganate	7722647		100	1		B	100 (45.4)
Potassium silver cyanide	506616	Argentate (1 -), bis(cyano-C) -, potassium	1*	4	P099	X	1 (0.454)
Pronamide	2395058 5	Benzamid e, 3,5-dichlo ro-N-(1,1-dimethyl-2--propynyl)--	1*	4	U192	D	5000 (2270)

Propanal, 2-methyl-2-(methylthio)-, O-[(methylamino)carbonyl]oxime	116063	Aldicarb	1*	4	P070	X	1 (0.454)
1-Propanamine	107108	n-Propylamine	1*	4	U194	D	5000 (2270)
1-Propanamine, N-propyl-	142847	Dipropylamine	1*	4	U110	D	5000 (2270)
1-Propanamine, N-nitroso-N-propyl-	621647	Di-n-propylnitrosamine	1*	2,4	U111	A	10 (4.54)
Propane, 1,2-dibromo-3-chloro-	96128	1,2-Dibromo-3-chloropropane	1*	4	U066	X	1 (0.454)
Propane, 2-nitro-	79469	2-Nitropropane	1*	4	U171	A	10 (4.54)
1,3-Propane sultone	1120714	1,2-Oxathiolane, 2,2-dioxide	1*	4	U193	A	10 (4.54)
Propane, 1,2-dichloro-	78875	Propylene dichloride<cr>1,2-Dichloropropane	5000	1,2,4	U083	C	1000 (454)
Propanedinitrile	109773	Malononitrile	1*	4	U149	C	1000 (454)
Propanenitrile	107120	Ethyl cyanide	1*	4	P101	A	10 (4.54)
Propanenitrile, 3-chloro-	542767	3-Chloropropionitrile	1*	4	P027	C	1000 (454)
Propanenitrile, 2-hydroxy-2-methyl-	75865	Acetone cyanohydrin<cr>2-Methyllactonitrile	10	1,4	P069	A	10 (4.54)
Propane, 2,2'-oxybis[2-chloro-	108601	Dichloroisopropyl ether	1*	2,4	U027	C	1000 (454)
1,2,3-Propanetriol, trinitrate-	55630	Nitroglycerine	1*	4	P081	A	10 (4.54)
1-Propanol, 2,3-dibromo-, phosphate (3:1)	126727	Tris(2,3-dibromopropyl) phosphate	1*	4	U235	A	10 (4.54)
1-Propanol, 2-methyl-	78831	Isobutyl alcohol	1*	4	U140	D	5000 (2270)
2-Propanone	67641	Acetone	1*	4	U002	D	5000 (2270)
2-Propanone, 1-bromo-	598312	Bromoacetone	1*	4	P017	C	1000 (454)
Propargite	2312358		10	1		A	10 (4.54)
Propargyl alcohol	107197	2-Propyn-1-ol	1*	4	P102	C	1000 (454)
2-Propenal	107028	Acrolein	1	1,2,4	P003	X	1 (0.454)
2-Propenamide	79061	Acrylamide	1*	4	U007	D	5000 (2270)
1-Propene, 1,1,2,3,3,3-hexachloro-	1888717	Hexachloropropene	1*	4	U243	C	1000 (454)

1-Propene, 1,3-dichloro-	542756	1,3-Dichloropropene	5000	1,2,4	U084	B	100 (45.4)
2-Propenenitrile	107131	Acrylonitrile	100	1,2,4	U009	B	100 (45.4)
2-Propenenitrile, 2-methyl-	126987	Methacrylonitrile	1*	4	U152	C	1000 (454)
2-Propenoic acid	79107	Acrylic acid	1*	4	U008	D	5000 (2270)
2-Propenoic acid, ethyl ester	140885	Ethyl acrylate	1*	4	U113	C	1000 (454)
2-Propenoic acid, 2-methyl-, ethyl ester	97632	Ethyl methacrylate	1*	4	U118	C	1000 (454)
2-Propenoic acid, 2-methyl-, methyl ester	80626	Methyl methacrylate	5000	1,4	U162	C	1000 (454)
2-Propen-1-ol	107186	Allyl alcohol	100	1,4	P005	B	100 (45.4)
Propionic acid	79094		5000	1		D	5000 (2270)
Propionic acid, 2-(2,4,5-trichloro phenoxy)-	93721	Silvex (2,4,5-TP) <cr>2,4,5--TP acid	100	1,4	U233	B	100 (45.4)
Propionic anhydride	123626		5000	1		D	5000 (2270)
n-Propylamine	107108	1-Propanamine	1*	4	U194	D	5000 (2270)
Propylene dichloride	78875	Propane, 1,2-dichloro-<cr>1, 2-Dichloropropane	5000	1,2,4	U083	C	1000 (454)
Propylene oxide	75569		5000	1		B	100 (45.4)
1,2-Propylenimine	75558	Aziridine, 2-methyl-	1*	4	P067	X	1 (0.454)
2-Propyn-1-ol	107197	Propargyl alcohol	1*	4	P102	C	1000 (454)
Pyrene	129000		1*	2		D	5000 (2270)
Pyrethrins	121299<cr>121211 <cr>8003347		1000	1		X	1 (0.545)
3,6-Pyridazinedione, 1,2-dihydro-	123331	Maleic hydrazide	1*	4	U148	D	5000 (2270)
4-Pyridinamine	504245	4-Aminopyridine	1*	4	P008	C	1000 (454)
Pyridine	110861		1*	4	U196	C	1000 (454)
Pyridine, 2-methyl-	109068	2-Picoline	1*	4	U191	D	5000 (2270)
Pyridine, 3-(1-methyl-2-pyrrolidinyl)-, (S)-	54115	Nicotine, & salts	1*	4	P075	B	100 (45.4)
2,4-(1H,3H)-Pyrimidinedione, 5-[bis(2-chloroethyl)amino]-	66751	Uracil mustard	1*	4	U237	A	10 (4.54)

4(1H)-Pyrimidinone, 2,3-dihydro-6-methyl-2-thioxo-	56042	Methylthiouracil	1*	4	U164	A	10 (4.54)
Pyrrolidine, 1-nitroso-	930552	N-Nitrosopyrrolidine	1*	4	U180	X	1 (0.454)
Quinoline	91225		1000	1		D	5000 (2270)
RADIONUCLIDES	N.A.		1*	3			
Reserpine	50555	Yohimban--16-carboxylic acid, 11,17-dimethoxy-18-[(3,4,5-trimethoxybenzoyl)oxy-, methyl ester (3beta, 16beta,17alpha,18beta,20alpha)-	1*	4	U200	D	5000 (2270)
Resorcinol	108463	1,3-Benzenediol	1000	1,4	U201	D	5000 (2270)
Saccharin and salts	81072	1,2-Benzisothiazol-3(2H)-one, 1,1-dioxide	1*	4	U202	B	100 (45.4)
Safrole	94597	1,3-Benzodioxole, 5-(2-propenyl)-	1*	4	U203	B	100 (45.4)
Selenious acid	7783008		1*	4	U204	A	10 (4.54)
Selenious acid, dithallium (1+) salt	12039520	Thallium selenite	1*	4	P114	C	1000 (454)
Selenium	7782492		1*	2		B	100 (45.4)
SELENIUM AND COMPOUNDS	N.A.		1*	2			**
Selenium dioxide	7446084	Selenium oxide	1000	1,4	U204	A	10 (4.54)
Selenium oxide	7446084	Selenium dioxide	1000	1,4	U204	A	10 (4.54)
Selenium sulfide	7488564	Selenium sulfide SeS2	1*	4	U205	A	10 (4.54)
Selenium sulfide SeS2	7488564	Selenium sulfide	1*	4	U205	A	10 (4.54)
Selenourea	630104		1*	4	P103	C	1000 (454)
L-Serine, diazoacetate (ester)	115026	Azaserine	1*	4	U015	X	1 (0.454)
Silver	7440224		1*	2		C	1000 (454)
SILVER AND COMPOUNDS	N.A.		1*	2			**

Silver cyanide	506649	Silver cyanide Ag (CN)	1*	4	P104	X	1 (0.454)
Silver cyanide Ag (CN)	506649	Silver cyanide	1*	4	P104	X	1 (0.454)
Silver nitrate	7761888		1	1		X	1 (0.454)
Silvex (2,4,5-TP)	93721	Propionic acid, 2-(2,4,5-trichlorophenoxy)-<cr>2,4,5-TP acid	100	1,4	U233	B	100 (45.4)
Sodium	7440235		1000	1		A	10 (4.54)
Sodium arsenate	7631892		1000	1		X	1 (0.454)
Sodium arsenite	7784465		1000	1		X	1 (0.454)
Sodium azide	26628228		1*	4	P105	C	1000 (454)
Sodium bichromate	10588019		1000	1		A	10 (4.54)
Sodium bifluoride	1333831		5000	1		B	100 (45.4)
Sodium bisulfite	7631905		5000	1		D	5000 (2270)
Sodium chromate	7775113		1000	1		A	10 (4.54)
Sodium cyanide	143339	Sodium cyanide Na (CN)	10	1,4	P106	A	10 (4.54)
Sodium cyanide Na (CN)	143339	Sodium cyanide	10	1,4	P106	A	10 (4.54)
Sodium dodecylbenzenesulfonate	25155300		1000	1		C	1000 (454)
Sodium fluoride	7681494		5000	1		C	1000 (454)
Sodium hydrosulfide	16721805		5000	1		D	5000 (2270)
Sodium hydroxide	1310732		1000	1		C	1000 (454)
Sodium hypochlorite	7681529<cr>10022705		100	1		B	100 (45.4)
Sodium methylate	124414		1000	1		C	1000 (454)
Sodium nitrite	7632000		100	1		B	100 (45.4)
Sodium phosphate, dibasic	7558794<cr>10039324<cr>10140655		5000	1		D	5000 (2270)
Sodium phosphate, tribasic	7601549<cr>7758294<cr>7785844<cr>10101890<cr>10124568<cr>10361894		5000	1		D	5000 (2270)
Sodium selenite	10102188<cr>7782823		1000	1		B	100 (45.4)

Streptozotocin	18883664	D-Glucose, 2-deoxy-2--[[(methylnitrosoamino)-carbonyl]amino]-<cr>Glucopyranose, 2-deoxy-2--(3-methyl--3-nitrosoureido)-	1*	4	U206	X	1 (0.454)
Strontium chromate	7789062		1000	1		A	10 (4.54)
Strychnidin-10-one	57249	Strychnine, & salts	10	1,4	P108	A	10 (4.54)
Strychnidin-10-one, 2,3-dimethoxy-	357573	Brucine	1*	4	P018	B	100 (45.4)
Strychnine, & salts	57249	Strychnidin-10-one	10	1,4	P108	A	10 (4.54)
Styrene	100425		1000	1		C	1000 (454)
Sulfur monochloride	12771083		1000	1		C	1000 (454)
Sulfur phosphide	1314803	Phosphorus pentasulfide <cr>Phosphorus sulfide	100	1,4	U189	B	100 (45.4)
Sulfuric acid	7664939 <cr>8014957		1000	1		C	1000 (454)
Sulfuric acid, dithallium (1+) salt	7446186 <cr>10031591	Thallium (I) sulfate	1000	1,4	P115	B	100 (45.4)
Sulfuric acid, dimethyl ester	77781	Dimethyl sulfate	1*	4	U103	B	100 (45.4)
2,4,5-T acid	93765	Acetic acid, (2,4,5-trichlorophenoxy)<cr>2,4,5-T	100	1,4	U232	C	1000 (454)
2,4,5-T amines	2008460 <cr>1319728<cr>3813147<cr>6369966<cr>6369977		100	1		D	5000 (2270)
2,4,5-T esters	93798 <cr>1928478 <cr>2545597 <cr>25168154 <cr>61792072		100	1		C	1000 (454)
2,4,5-T salts	13560991		100	1		C	1000 (454)

2,4,5-T	93765	Acetic acid, (2,4,5-trichlorophenoxy)<cr>2,4,5-T acid	100	1,4	U232	C	1000 (454)
TDE	72548	Benzene, 1,1'-(2,2-dichloroethylidene)bis[4-chloro- - DDD 4,4' DDD	1	1,2,4	U060	X	1 (0.454)
1,2,4,5-Tetrachlorobenzene	95943	Benzene, 1,2,4,5-tetrachloro-	1*	4	U207	D	5000 (2270)
2,3,7,8-Tetrachlorodibenzo-p-dioxin (TCDD)	1746016		1*	2		X	1 (0.454)
1,1,1,2-Tetrachloroethane	630206	Ethane, 1,1,1,2-tetrachloro-	1*	4	U208	B	100 (45.4)
1,1,2,2-Tetrachloroethane	79345	Ethane, 1,1,2,2-tetrachloro-	1*	2,4	U209	B	100 (45.4)
Tetrachloroethene	127184	Ethene, tetrachloro--<cr>Perchloroethylene <cr>Tetrachloroethylene	1*	2,4	U210	B	100 (45.4)
Tetrachloroethylene	127184	Ethene, tetrachloro-- <cr>Perchloroethylene Tetrachloroethene	1*	2,4	U210	B	100 (45.4)
2,3,4,6-Tetrachlorophenol	58902	Phenol, 2,3,4,6-tetrachloro-	1*	4	U212	A	10 (4.54)
Tetraethyl lead	78002	Plumbane, tetraethyl-	100	1,4	P110	A	10 (4.54)
Tetraethyl pyrophosphate	107493	Diphosphoric acid, tetraethyl ester	100	1,4	P111	A	10 (4.54)
Tetraethyldithiopyrophosphate	3689245	Thiodiphosphoric acid, tetraethyl ester	1*	4	P109	B	100 (45.4)
Tetrahydrofuran	109999	Furan, tetrahydro--	1*	4	U213	C	1000 (454)
Tetranitromethane	509148	Methane, tetranitro-	1*	4	P112	A	10 (4.54)
Tetraphosphoric acid, hexaethyl ester	757584	Hexaethyl tetraphosphoate	1*	4	P062	B	100 (45.4)

Thallic oxide	1314325	Thallium oxide Tl203	1*	4	P113	B	100 (45.4)
Thallium	7440280		1*	2		C	1000 (454)
Thallium and compounds	N.A.		1*	2			**
Thallium (I) acetate	563688	Acetic acid, thallium(1+) salt	1*	4	U214	B	100 (45.4)
Thallium (I) carbonate	6533739	Carbonic acid, dithallium(1+) salt	1*	4	U215	B	100 (45.4)
Thallium (I) chloride	7791120	Thallium chloride TlCl	1*	4	U216	B	100 (45.4)
Thallium chloride TlCl	7791120	Thallium(I) chloride	1*	4	U216	B	100 (45.4)
Thallium (I) nitrate	10102451	Nitric acid, thallium (1+) salt	1*	4	U217	B	100 (45.4)
Thallium oxide Tl203	1314325	Thallic oxide	1*	4	P113	B	100 (45.4)
Thallium selenite	12039520	Selenious acid, dithallium(1+) salt	1*	4	P114	C	1000 (454)
Thallium (I) sulfate	7446186 <cr>10031591	Sulfuric acid, dithallium(1+) salt	1000	1,4	P115	B	100 (45.4)
Thioacetamide	62555	Ethanethioamide	1*	4	U218	A	10 (4.54)
Thiodiphosphoric acid, tetraethyl ester	3689245	Tetraethyl dithiopyrophosphate	1*	4	P109	B	100 (45.4)
Thiofanox	39196184	2-Butanone, 3,3-dimethyl-1-(methylthio)-, O[(methylamino)carbonyl) oxime	1*	4	P045	B	100 (45.4)
Thioimidodicarbonic diamide [(H2N)C(S)] 2NH	541537	Dithiobiuret	1*	4	P049	B	100 (45.4)
Thiomethanol	74931	Methanethiol <cr>Methylmercaptan	100	1,4	U153	B	100 (45.4)
Thioperoxydicarbonic diamide [(H2N)C(S)] 2S2, tetramethyl-	137268	Thiram	1*	4	U244	A	10 (4.54)
Thiophenol	108985	Benzenethiol	1*	4	P014	B	100 (45.4)

Thiosemicarbazide	79196	Hydrazine carbothio amide	1*	4	P116	B	100 (45.4)
Thiourea	62566		1*	4	U219	A	10 (4.54)
Thiourea, (2-chlorophenyl)-	5344821	1-(o-Chlorophenyl)thiourea	1*	4	P026	B	100 (45.4)
Thiourea, 1-naphthalenyl-	86884	alpha-Naphthylthio urea	1*	4	P072	B	100 (45.4)
Thiourea, phenyl-	103855	Phenylthio urea	1*	4	P093	B	100 (45.4)
Thiram	137268	Thioperoxydicarbonic diamide<cr>[(H2N)C(S)] 2S2, tetramethyl-	1*	4	U244	A	10 (4.54)
Toluene	108883	Benzene, methyl-	1000	1,2,4	U220	C	1000 (454)
Toluenediamine	95807 <cr>496720 <cr>823405 <cr>25376458	Benzenediamine, ar-methyl-	1*	4	U221	A	10 (4.54)
Toluene diisocyanate	584849 <cr>91087 <cr>26471625	Benzene, 1,3-diisocyanatomethyl-	1*	4	U223	B	100 (45.4)
o-Toluidine	95534	Benzenamine, 2-methyl-	1*	4	U328	B	100 (45.4)
p-Toluidine	106490	Benzenamine, 4-methyl-	1*	4	U353	B	100 (45.4)
o-Toluidine hydrochloride	636215	Benzenamine, 2-methyl-, hydrochloride	1*	4	U222	B	100 (45.4)
Toxaphene	8001352	Camphene, octachloro-	1*	1,2,4,	P123	X	1 (0.454)
2,4,5-TP acid	93721	Propionic acid, 2-(2,4,5-trichlorophenoxy)-<cr>Silvex (2,4,5-TP)	100	1,4	U233	B	100 (45.4)
2,4,5-TP esters	32534955		100	1		B	100 (45.4)
1H-1,2,4-Triazol-3-amine	61825	Amitrole	1*	4	U011	A	10 (4.54)
Trichlorfon	52686		1000	1		B	100 (45.4)
1,2,4-Trichlorobenzene	120821		1*	2		B	100 (45.4)

1,1,1-Trichloroethane	71556	Ethane, 1,1,1-trichloro-<cr>Methyl chloroform	1*	2,4	U226	C	1000 (454)
1,1,2-Trichloroethane	79005	Ethane, 1,1,2-trichloro-	1*	2,4	U227	B	100 (45.4)
Trichloroethene	79016	Ethene, trichloro-<cr>Trichloroethylene	1000	1,2,4	U228	B	100 (45.4)
Trichloroethylene	79016	Ethene, trichloro-<cr>Trichloroethene	1000	1,2,4	U228	B	100 (45.4)
Trichloromethane sulfenyl chloride	594423	Methanesulfenyl chloride, trichloro-	1*	4	P118	B	100 (45.4)
Trichloromonofluoromethane	75694	Methane, trichlorofluoro-	1*	4	U121	D	5000 (2270)
Trichlorophenol	25167822		10	1		A	10 (4.54)
2,3,4-Trichlorophenol	15950660						
2,3,5-Trichlorophenol	933788						
2,3,6-Trichlorophenol	933755						
2,4,5-Trichlorophenol	95954	Phenol, 2,4,5-trichloro-	10*	1,4	U230	A	10 (4.54)
2,4,6-Trichlorophenol	88062	Phenol, 2,4,6-trichloro-	10*	1,2,4	U231	A	10 (4.54)
3,4,5-Trichlorophenol	609198						
2,4,5-Trichlorophenol	95954	Phenol, 2,4,5-trichloro-	10*	1,4	U230	A	10 (4.54)
2,4,6-Trichlorophenol	88062	Phenol, 2,4,6-trichloro-	10	1,2,4	U231	A	10 (4.54)
Triethanolamine dodecylbenzenesulfonate	27323417		1000	1		C	1000 (454)
Triethylamine	121448		5000	1		D	5000 (2270)
Trimethylamine	75503		1000	1		B	100 (45.4)
1,3,5-Trinitrobenzene	99354	Benzene, 1,3,5-trinitro-	1*	4	U234	A	10 (4.54)
1,3,5-Trioxane, 2,4,6-trimethyl-	123637	Paraldehyde	1*	4	U182	C	1000 (454)

Tris(2,3-dibromo propyl) phosphate	126727	1-Propanol, 2,3-dibromo-, phosphate [(3:1)	1*	4	U235	A	10 (4.54)
Trypan blue	72571	2,7-Naphthalenedisulfonic acid, 3,3'-3,3'-dimethyl-(1,1'-biphenyl)-4,4'-diyl)-bis(azo)]bis(5-amino-4-hydroxy)-tetrasodium salt	1*	4	U236	A	10 (4.54)
Unlisted Hazardous Wastes Characteristic of Corrosivity	N.A.		1*	4	D002	B	100 (45.4)
Unlisted Hazardous Wastes Characteristics:	N.A.		1*	4			
Characteristic of Toxicity:							
Arsenic (D004)	N.A.		*1	4	D004	X	1(0.454)
Barium (D005)	N.A.		*1	4	D005	C	1,000(454)
Benzene (D018)	N.A.		1000	1, 2, 3, 4	D018	A	10(4.54)
Cadmium (D006)	N.A.		*1	4	D006	A	10(4.54)
Carbon tetrachloride(D019)	N.A.		5,000	1, 2, 4	D019	A	10(4.54)
Chlordane (D020)	N.A.		1	1, 2, 4	D020	X	1(0.454)
Chlorobenzene (D021)	N.A.		100	1, 2, 4	D021	B	100(45.4)
Chloroform(D022)	N.A.		5,000	1, 2, 4	D022	A	10(4.54)
Chromium (D007)	N.A.		*1	4	D007	A	10(4.54)
o-Cresol (D023)	N.A.		1,000	1, 4	D023	C	1,000(454)
m-Cresol (D024)	N.A.		1,000	1, 4	D024	C	1,000(454)
p-Cresol (D025)	N.A.		1,000	1, 4	D025	C	1,000(454)
Cresol (D026)	N.A.		1,000	1, 4	D026	C	1,000(454)
2,4-D (D016)	N.A.		100	1, 4	D016	B	100(45.4)
1,4-Dichlorobenzene (D027)	N.A.		100	1, 2, 4	D027	B	100(45.4)

1,2-Dichloroethane (D028)	N.A.		5,000	1, 2, 4	D028	B	100(45.4)
1,1-Dichloroethylene (D029)	N.A.		5,000	1, 2, 4	D029	B	100(45.4)
2,4-Dinitrotoluene (D030)	N.A.		1,000	1, 2, 4	D030	A	10(4.54)
Endrin (D012)	N.A.		1	1, 4	D012	X	1(0.454)
Heptachlor (and epoxide) (D031)	N.A.		1	1, 2, 4	D031	X	1(0.454)
Hexachlorobenzene (D032)	N.A.		*1	2, 4	D032	A	10(4.54)
Hexachlorobutadiene (D033)	N.A.		*1	2, 4	D033	X	1(0.454)
Hexachloroethane (D034)	N.A.		*1	2, 4	D034	B	100(45.4)
Lead (D008)	N.A.		1*	4	D008	A	10 (4.54)
Lindane (D013)	N.A.		1	1, 4	D013	X	1(0.454)
Mercury (D009)	N.A.		*1	4	D009	X	1(0.454)
Methoxychlor (D014)	N.A.		1	1, 4	D014	X	1(0.454)
Methyl ethyl ketone (D035)	N.A.		*1	4	D035	D	5,000(2270)
Nitrobenzene (D036)	N.A.		1,000	1, 2, 4	D036	C	1,000(454)
Pentachlorophenol (D037)	N.A.		10	1, 2, 4	D037	A	10(4.54)
Pyridine (D038)	N.A.		*1	4	D038	C	1,000(454)
Selenium (D010)	N.A.		*1	4	D010	A	10(4.54)
Silver (D011)	N.A.		*1	4	D011	X	1(0.454)
Tetrachloroethylene (D039)	N.A.		*1	2, 4	D039	B	100(45.4)
Toxaphene (D015)	N.A.		1	1, 4	D015	X	1(0.454)
Trichloroethylene (D040)	N.A.		1000	1, 2, 4	D040	B	100(45.4)
2,4,5-Trichlorophenol (D041)	N.A.		10	1, 4	D041	A	10(4.54)
2,4,6-Trichlorophenol (D042)	N.A.		10	1, 2, 4	D042	A	10(4.54)
2,4,5-TP (D017)	N.A.		100	1, 4	D017	B	100(45.4)
Vinyl chloride (D043)	N.A.		*1	2, 3, 4	D043	X	1(0.454)
Unlisted Hazardous Wastes Characteristic of Ignitability	N.A.		1*	4	D001	B	100 (45.4)

Unlisted Hazardous Wastes Characteristic of Reactivity	N.A.		1*	4	D003	B	100 (45.4)
Uracil mustard	66751	2,4-(1H,3H)-Pyrimidinedione, 5-[bis(2-chloroethyl)amino]-	1*	4	U237	A	10 (4.54)
Uranyl acetate	541093		5000	1		B	100 (45.4)
Uranyl nitrate	10102064<cr>36478769		5000	1		B	100 (45.4)
Urea, N-ethyl-N-nitroso--	759739	N-Nitroso-N-ethylurea	1*	4	U176	X	1 (0.454)
Urea, N-methyl-N-nitroso	684935	N-Nitroso-N-methyl urea	1*	4	U177	X	1 (0.454)
Vanadic acid, ammonium salt	7803556	Ammonium vanadate	1*	4	P119	C	1000 (454)
Vanadium oxide V205	1314621	Vanadium pentoxide	1000	1,4	P120	C	1000 (454)
Vanadium pentoxide	1314621	Vanadium oxide V205	1000	1,4	P120	C	1000 (454)
Vanadyl sulfate	27774136		1000	1		C	1000 (454)
Vinyl chloride	75014	Ethene, chloro-	1*	2,3,4	U043	X	1 (0.454)
Vinyl acetate	108054	Vinyl acetate monomer	1000	1		D	5000 (2270)
Vinyl acetate monomer	108054	Vinyl acetate	1000	1		D	5000 (2270)
Vinylamine, N-methyl-N-nitroso-	4549400	N-Nitrosomethylvinylamine	1*	4	P084	A	10 (4.54)
Vinylidene chloride	75354	Ethene, 1,1-dichloro-<cr>1,1-Dichloroethylene	5000	1,2,4	U078	B	100 (45.4)
Warfarin, & salts, when present at concentrations greater than 0.3%	81812	2H-1-Benzopyran-2--one, 4-hydroxy-3-(3-oxo-1-phenyl-butyl)-, & salts, when present at concentrations greater than 0.3%	1*	4	P001	B	100 (45.4)
Xylene (mixed)	1330207	Benzene, dimethyl	1000	1,4	U239	C	1000 (454)

m-Benzene, dimethyl	108383	m-Xylene					
o-Benzene, dimethyl	95476	o-Xylene					
p-Benzene, dimethyl	106423	p-Xylene					
Xylenol	1300716		1000	1		C	1000 (454)
Yohimban-16-carboxylic acid,11,17-dimethoxy-18-[(3,4,5-trimethoxybenzoyl)oxy]-, methyl ester (3beta,16beta,17alpha,<cr>18beta, 20alpha)-	50555	Reserpine	1*	4	U200	D	5000 (2270)
Zinc	7440666		1*	2		C	1000 (454)
ZINC AND COMPOUNDS	N.A.		1*	2			**
Zinc acetate	557346		1000	1		C	1000 (454)
Zinc ammonium chloride	52628258<cr>14639975<cr>14639986		5000	1		C	1000 (454)
Zinc borate	1332076		1000	1		C	1000 (454)
Zinc bromide	7699458		5000	1		C	1000 (454)
Zinc carbonate	3486359		1000	1		C	1000 (454)
Zinc chloride	7646857		5000	1		C	1000 (454)
Zinc cyanide	557211	Zinc cyanide Zn(CN)2	10	1,4	P121	A	10 (4.54)
Zinc cyanide Zn(CN)2	557211	Zinc cyanide	10	1,4	P121	A	10 (4.54)
Zinc fluoride	7783495		1000	1		C	1000 (454)
Zinc formate	557415		1000	1		C	1000 (454)
Zinc hydrosulfite	7779864		1000	1		C	1000 (454)
Zinc nitrate	7779886		5000	1		C	1000 (454)
Zinc phenosulfonate	127822		5000	1		D	5000 (2270)
Zinc phosphide	1314847	Zinc phosphide Zn3P2, when present at concentrations greater than 10%	1000	1,4	P122	B	100 (45.4)

Zinc phosphide Zn3P2, when present at concentrations greater than 10%	1314847	Zinc phosphide	1000	1,4	P122	B	100 (45.4)
Zinc silicofluoride	16871719		5000	1		D	5000 (2270)
Zinc sulfate	7733020		1000	1		C	1000 (454)
Zirconium nitrate	13746899		5000	1		D	5000 (2270)
Zirconium potassium fluoride	16923958		5000	1		C	1000 (454)
Zirconium sulfate	14644612		5000	1		D	5000 (2270)
Zirconium tetrachloride	10026116		5000	1		D	5000 (2270)
F001			1*	4	F001	A	10 (4.54)
The following spent halogenated solvents used in degreasing; all spent solvent mixtures/blends used in degreasing containing, before use, a total of ten percent or more (by volume) of one or more of the above halogenated solvents or those solvents listed in F002, F004, and F005; and still bottoms from the recovery of these spent solvents and spent solvent mixtures.							
(a) Tetrachloroethylene	127184		1*	2,4	U210	B	100 (45.4)
(b) Trichloroethylene	79016		1000	1,2,4	U228	B	100 (45.4)
(c) Methylene chloride	75092		1*	2,4	U080	C	1000 (454)
(d) 1,1,1-Trichloroethane	71556		1*	2,4	U226	C	1000 (454)
(e) Carbon tetrachloride	56235		5000	1,2,4	U211	A	10 (4.54)
(f) Chlorinated fluorocarbons	N.A.					D	5000 (2270)
F002			1*	4	F002	A	10 (4.54)

The following spent halogenated solvents; all spent solvent mixtures/blends containing, before use, a total of ten percent or more (by volume) of one or more of the above halogenated solvents or those listed in F001, F004, or F005; and still bottoms from the recovery of these spent solvents and spent solvent mixtures.							
(a) Tetrachloroethylene	127184		1*	2,4	U210	B	100 (45.4)
(b) Methylene chloride	75092		1*	2,4	U080	C	1000 (454)
(c) Trichloroethylene	79016		1000	1,2,4	U228	B	100 (45.4)
(d) 1,1,1-Trichloroethane	71556		1*	2,4	U226	C	1000 (454)
(e) Chlorobenzene	108907		100	1,2,4	U037	B	100 (45.4)
(f) 1,1,2-Trichloro-1, 2,2-trifluoroethane	76131					D	5000 (2270)
(g) o-Dischlorobenzene	95501		100	1,2,4	U070	B	100 (45.4)
(h) Trichlorofluoromethane	75694		1*	4	U121	D	5000 (2270)
(i) 1,1,2-Thrichloroethane	79005		1*	2,4	U227	B	100 (45.4)
F003			1*	4	F003	B	100 (45.4)
The following spent non-halogenated solvents and the still bottoms from the recovery of these solvents:							
(a) Xylene	1330207					C	1000 (454)
(b) Acetone	67641					D	5000 (2270)
(c) Ethyl acetate	141786					D	5000 (2270)

(d) Ethylbenzene	100414					C	1000 (454)
(e) Ethyl ether	60297					B	100 (45.4)
(f) Methyl isobutyl ketone	108101					D	5000 (2270)
(g) n-Butyl alcohol	71363					D	5000 (2270)
(h) Cyclohexanone	108941					D	5000 (2270)
(i) Methanol	67561					D	5000 (2270)
F004			1*	4	F004	C	1000 (454)
The following spent non-halogenated solvents and the still bottoms from the recovery of these solvents:							
(a) Cresols/Cresylic acid	1319773		1000	1,4	U052	C	1000 (454)
(b) Nitrobenzene	98953		1000	1,2,4	U169	C	1000 (454)
F005			1*	4	F005	B	100 (45.4)
The following spent non-halogenated solvents and the still bottoms from the recovery of these solvents:							
(a) Toluene	108883		1000	1,2,4	U220	C	1000 (454)
(b) Methyl ethyl ketone	78933		1*	4	U159	D	5000 (2270)
(c) Carbon disulfide	75150		5000	1,4	P022	B	100 (45.4)
(d) Isobutanol	78831		1*	4	U140	D	5000 (2270)
(e) Pyridine	110861		1*	4	U196	C	1000 (454)
F006			1*	4	F006	A	10 (4.54)

Wastewater treatment sludges from electroplating operations except from the following processes: (1) sulfuric acid anodizing of aluminum, (2) tin plating on carbon steel, (3) zinc plating (segregated basis) on carbon steel, (4) aluminum or zinc-aluminum plating on carbon steel, (5) cleaning/stripping associated with tin, zinc and aluminum plating on carbon steel, and (6) chemical etching and milling of aluminum.							
F007			1*	4	F007	A	10 (4.54)
Spent cyanide plating bath solutions from electroplating operations.							
F008			1*	4	F008	A	10 (4.54)
Plating bath residues from the bottom of plating baths from electroplating operations where cyanides are used in the process.							
F009			1*	4	F009	A	10 (4.54)
Spent stripping and cleaning bath solutions from electroplating operations where cyanides are used in the process.							
F010			1*	4	F010	A	10 (4.54)

Quenching bath residues from oil baths from metal heat treating operations where cyanides are used in the process.							
F011			1*	4	F011	A	10 (4.54)
Spent cyanide solution from salt bath pot cleaning from metal heat treating operations.							
F012			1*	4	F012	A	10 (4.54)
Quenching wastewater treatment sludges from metal heat treating operations where cyanides are used in the process.							
F019			1	4	F019	A	10 (4.54)
Wastewater treatment sludges from the chemical conversion coating of aluminum except from zirconium phosphating in aluminum can washing when such phosphating is an exclusive conversion coating process							
F020			1*	4	F020	X	1 (0.454)

Wastes (except wastewater and spent carbon from hydrogen chloride purification) from the production or manufacturing use (as a reactant, chemical intermediate, or component in a formulating process) of tri-or-tetrachlorophenol, or of intermediates used to produce their pesticide derivatives. (This listing does not include wastes from the production of hexachlorophene from highly purified 2,4,5-trichlorophenol.)							
F021			1*	4	F021	X	1 (0.454)
Wastes (except wastewater and spent carbon from hydrogen chloride purification) from the production or manufacturing use (as a reactant, chemical intermediate, or component in a formulating process) of pentachlorophenol, or of intermediates used to produce its derivatives.							
F022			1*	4	F022	X	1 (0.454)

Wastes (except wastewater and spent carbon from hydrogen chloride purification) from the manufacturing use (as a reactant, chemical intermediate, or component in a formulating process) of tetra-, penta-, or hexachlorobenzenes under alkaline conditions.							
F023			1*	4	F023	X	1 (0.454)
Wastes (except wastewater and spent carbon from hydrogen chloride purification) from the production of materials on equipment previously used for the production or manufacturing use (as a reactant, chemical intermediate, or component in a formulating process) of tri- and tetrachlorophenols. (This listing does not include wastes from equipment used only for the production or use of hexa-chlorophene from highly purified 2,4,5-tri-chlorophenol.)							
F024			1*	4	F024	X	1 (0.454)

Wastes, including but not limited to distillation residues, heavy ends, tars, and reactor cleanout wastes, from the production of chlorinated aliphatic hydrocarbons, having carbon content from one to five, utilizing free radical catalyzed processes. (This listing does not include light ends, spent filters and filter aids, spent dessicants(sic), wastewater, wastewater treatment sludges, spent catalysts, and wastes listed in Section 261.32.)							
F025			1*	4	F025	X	F1F11 (0.454)
Condensed light ends, spent filters and filter aids, and spent desiccant wastes from the producton of certain chlorinated aliphatic hydrocarbons, by free radical catalyzed processes. These chlorinated aliphatic hydrocarbons are those having carbon chain lengths ranging from one to and includng five, with varying amounts and positions of chlorine substitution.							
F026			1*	4	F026	X	1 (0.454)

Wastes (except wastewater and spent carbon from hydrogen chloride purification) from the production of materials on equipment previously used for the manufacturing use (as a reactant, chemical intermediate, or component in a formulating process) of tetra-, penta-, or hexachlorobenzene under alkaline conditions.							
F027			1*	4	F027	X	1 (0.454)
Discarded unused formulations containing tri-, tetra-, or pentachlorophenol or discarded unused formulations containing compounds derived from these chlorophenols. (This listing does not include formulations containing hexachlorophene synthesized from prepurified 2,4,5-tri-chlorophenol as the sole component.)							
F028			1*	4	F028	X	1 (0.454)
Residues resulting from the incineration or thermal treatment of soil contaminated with EPA Hazardous Waste Nos. F020, F021, F022, F023, F026, and F027.							

F032			1*	4	F032	X	1(0.454)
Wastewaters (except those that have not come into contact with process contaminants), process residuals, preservative drippage, and spent formulations from wood preserving processes generated at plants that currently use or have previously used chlorophenolic formulations (except potentially cross-contaminated wastes that have had the F032 waste code deleted in accordance with 261.35 of this chapter or potentially cross-contaminated wastes that are otherwise currently regulated as hazardous wastes (i.e., F034 or F035), and where the generator does not resume or initiate use of chlorophenolic formulations). This listing does not include K001 bottom sediment sludge from the treatment of wastewater from wood preserving processes that use creosote and/or pentachlorophenol.							
F034			1*	4	F034	X	1(0.454)

Wastewaters (except those that have not come into contact with process contaminants), process residuals, preservative drippage, and spent formulations from wood preserving processes generated at plants that use cresote formulations. This listing does not include K001 bottom sediment sludge from the treatment of wastewater from wood preserving processes that use creosote and/or pentachlorophenol.							
F035			1*	4	F035	X	1(0.454)
Wastewaters (except those that have not come into contact with process contaminants), process residuals, preservative drippage, and spent formulations from wood preserving processes generated at plants that use inorganic preservatives containing arsenic or chromium. This listing does not include K001 bottom sediment sludge from the treatment of wastewater from wood preserving processes that use creosote and/or pentachlorophenol.							
F037			1*	4	F037	X	1 (0.454)

Petroleum refinery primary oil/water/solids separation sludge-Any sludge generated from the gravitational separation of oil/water/solids during the storage or treatment of process wastewaters and oily cooling wastewaters from petroleum refineries. Such sludges include, but are not limited to, those generated in: oil/water/solids separators; tanks and impoundments; ditches and other conveyances; sumps; and stormwater units receiving dry weather flow. Sludge generated in stormwater units that do not receive dry weather flow, sludges generated from non-contact once-through cooling waters segregated for treatment from other process or oily cooling waters, sludges generated in aggressive biological treatment units as defined in 261.31(b)(2) (including sludges generated in one or more additional units after wastewaters have been treated in aggressive biological treatment units) and K051 wastes are not included in this listing.							

F038			1*	4	F038	X	1 (0.454)
Petroleum refinery secondary (emulsified) oil/water/solids separation sludge-Any sludge and/or float generated from the physical and/or chemical separation of oil/water/solids in process wastewaters and oily cooling wastewaters from petroleum refineries. Such wastes include, but are not limited to, all sludges and floats generated in: induced air flotation (IAF) units, tanks and impoundments, and all sludges generated in DAF units. Sludges generated in stormwater units that do not receive dry weather flow, sludges generated from once-through non-contact cooling waters segregated for treatment from other process or oil cooling wastes, sludges and floats generated in aggressive biological treatment units as defined in 261.31(b)(2) (including sludges and floats generated in one or more additional units after wastewaters have been treated in aggressive biological treatment units)							

and F037, K048, and K051 wastes are not included in this listing.							
K001			1*	4	K001	X	1 (0.454)
Bottom sediment sludge from the treatment of wastewaters from wood preserving processes that use creosote and/or pentachlorophen ol.							
K002			1*	4	K002	A	10 (4.54)
Wastewater treatment sludge from the production of chrome yellow and orange pigments.							
K003			1*	4	K003	A	10 (4.54)
Wastewater treatment sludge from the production of molybdate orange pigments.							
K004			1*	4	K004	A	10 (4.54)
Wastewater treatment sludge from the production of zinc yellow pigments.							
K005			1*	4	K005	A	10 (4.54)
Wastewater treatment sludge from the production of chrome green pigments.							
K006			1*	4	K006	A	10 (4.54)

Wastewater treatment sludge from the production of chrome oxide green pigments (anhydrous and hydrated).							
K007			1*	4	K007	A	10 (4.54)
Wastewater treatment sludge from the production of iron blue pigments.							
K008			1*	4	K008	A	10 (4.54)
Oven residue from the production of chrome oxide green pigments.							
K009			1*	4	K009	A	10 (4.54)
Distillation bottoms from the production of acetaldehyde from ethylene.							
K010			1*	4	K010	A	10 (4.54)
Distillation side cuts from the production of acetaldehyde from ethylene.							
K011			1*	4	K011	A	10 (4.54)
Bottom stream from the wastewater stripper in the production of acrylonitrile.							
K013			1*	4	K013	A	10 (4.54)

Bottom stream from the acetonitrile column in the production of acrylonitrile.							
K014			1*	4	K014	D	5000 (2270)
Bottoms from the acetonitrile purification column in the production of acrylonitrile.							
K015			1*	4	K015	A	10 (4.54)
Still bottoms from the distillation of benzyl chloride.							
K016			1*	4	K016	X	1 (0.454)
Heavy ends or distillation residues from the production of carbon tetrachloride.							
K017			1*	4	K017	A	10 (4.54)
Heavy ends (still bottoms) from the purification column in the production of epi-chlorohydrin.							
K018			1*	4	K018	X	1 (0.454)
Heavy ends from the fractionation column in ethyl chloride production.							
K019			1*	4	K019	X	1 (0.454)

Heavy ends from the distillation of ethylene dichloride in ethylene dichloride production.							
K020			1*	4	K020	X	1 (0.454)
Heavy ends from the distillation of vinyl chloride in vinyl chloride monomer production.							
K021			1*	4	K021	A	10 (4.54)
Aqueous spent antimony catalyst waste from fluoromethanes production.							
K022			1*	4	K022	X	1 (0.454)
Distillation bottom tars from the production of phenol/acetone from cumene.							
K023			1*	4	K023	D	5000 (2270)
Distillation light ends from the production of phthalic anhydride from naphthalene.							
K024			1*	4	K024	D	5000 (2270)
Distillation bottoms from the production of phthalic anhydride from naphthalene.							
K025			1*	4	K025	A	10 (4.54)

Distillation bottoms from the production of nitrobenzene by the nitration of benzene.							
K026			1*	4	K026	C	1000 (454)
Stripping still tails from the production of methyl ethyl pyridines.							
K027			1*	4	K027	A	10 (4.54)
Centrifuge and distillation residues from toluene diisocyanate production.							
K028			1*	4	K028	X	1 (0.454)
Spent catalyst from the hydrochlorinator reactor in the production of 1,1,1-trichloroeth ane.							
K029			1*	4	K029	X	1 (0.454)
Waste from the product steam stripper in the production of 1,1,1-trichloroeth ane.							
K030			1*	4	K030	X	1 (0.454)
Column bottoms or heavy ends from the combined production of trichloroethylene and perchloroethylene .							
K031			1*	4	K031	X	1 (0.454)

By-product salts generated in the production of MSMA and cacodylic acid.							
K032			1*	4	K032	A	10 (4.54)
Wastewater treatment sludge from the production of chlordane.							
K033			1*	4	K033	A	10 (4.54)
Wastewater and scrub water from the chlorination of cyclopentadiene in the production of chlordane.							
K034			1*	4	K034	A	10 (4.54)
Filter solids from the filtration of hexachlorocyclop entadiene in the production of chlordane.							
K035			1*	4	K035	X	1 (0.454)
Wastewater treatment sludges generated in the production of creosote.							
K036			1*	4	K036	X	1 (0.454)
Still bottoms from toluene reclamation distillation in the production of disulfoton.							
K037			1*	4	K037	X	1 (0.454)

Wastewater treatment sludges from the production of disulfoton.							
K038			1*	4	K038	A	10 (4.54)
Wastewater from the washing and stripping of phorate production.							
K039			1*	4	K039	A	10 (4.54)
Filter cake from the filtration of diethylphosphoro dithioic acid in the production of phorate.							
K040			1*	4	K040	A	10 (4.54)
Wastewater treatment sludge from the production of phorate.							
K041			1*	4	K041	X	1 (0.454)
Wastewater treatment sludge from the production of toxaphene.							
K042			1*	4	K042	A	10 (4.54)
Heavy ends or distillation residues from the distillation of tetrachlorobenzen e in the production of 2,4,5-T.							
K043			1*	4	K043	A	10 (4.54)
2,6-Dichlorophen ol waste from the production of 2,4-D.							

K044			1*	4	K044	A	10 (4.54)
Wastewater treatment sludges from the manufacturing and processing of explosives.							
K045			1*	4	K045	A	10 (4.54)
Spent carbon from the treatment of wastewater containing explosives.							
K046			1*	4	K046	A	10 (4.54)
Wastewater treatment sludges from the manufacturing, formulation and loading of lead-based initiating compounds.							
K047			1*	4	K047	A	10 (4.54)
Pink/red water from TNT operations.							
K048			1*	4	K048	A	10 (4.54)
Dissolved air flotation (DAF) float from the petroleum refining industry.							
K049			1*	4	K049	A	10 (4.54)
Stop oil emulsion solids from the petroleum refining industry.							
K050			1*	4	K050	A	10 (4.54)

Heat exchanger bundle cleaning sludge from the petroleum refining industry.							
K051			1*	4	K051	A	10 (4.54)
API separator sludge from the petroleum refining industry.							
K052			1*	4	K052	A	10 (4.54)
Tank bottoms (leaded) from the petroleum refining industry.							
K060			1*	4	K060	X	1 (0.454)
Ammonia still lime sludge from coking operations.							
K061			1*	4	K061	A	10 (4.54)
Emission control dust/sludge from the primary production of steel in electric furnaces.							
K062			1*	4	K062	A	10 (4.54)
Spent pickle liquor generated by steel finishing operations of factilties within the iron and steel industry (SIC Codes 331 and 332).							
K064			1*	4	K064	A	10 (4.54)

Acid plant blowdown slurry/sludge resulting from thickening of blowdown slurry from primary copper production.							
K065			1*	4	K065	A	10 (4.54)
Surface impoundment solids contained in and dredged from surface impoundments at primary lead smelting facilities.							
K066			1*	4	K066	A	10 (4.54)
Sludge from treatment of process wastewater and/or acid plant blowdown from primary zinc production.							
K069			1*	4	K069	A	10 (4.54)
Emission control dust/sludge from secondary lead smelting.							
K071			1*	4	K071	X	1 (0.454)
Brine purification muds from the mercury cell process in chlorine production, where separately prepurified brine is not used.							
K073			1*	4	K073	A	10 (4.54)

Chlorinated hydrocarbon waste from the purification step of the diaphragm cell process using graphite anodes in chlorine production.							
K083			1*	4	K083	B	100 (45.4)
Distillation bottoms from aniline extraction.							
K084			1*	4	K084	X	1 (0.454)
Wastewater treatment sludges generated during the production of veterinary pharmaceuticals from arsenic or organo-arsenic compounds.							
K085			1*	4	K085	A	10 (4.54)
Distillation or fractionation column bottoms from the production of chlorobenzenes.							
K086			1*	4	K086	A	10 (4.54)
Solvent washes and sludges, caustic washes and sludges, or water washes and sludges from cleaning tubs and equipment used in the formulation of ink from pigments, driers, soaps, and stabilizers containing chromium and lead.							
K087			1*	4	K087	B	100 (45.4)

Decanter tank tar sludge from coking operations.							
K088			1*	4	K088		
Spent potliners from primary aluminum reduction.							
K090			1*	4	K090		
Emission control dust or sludge from ferrochromiumsilicon production.							
K091			1	4	K091		
Emission control dust or sludge from ferrochromium production.							
K093			1*	4	K093	D	5000 (2270)
Distillation light ends from the production of phthalic anhydride from ortho-xylene.							
K094			1*	4	K094	D	5000 (2270)
Distillation bottoms from the production of phthalic anhydride from ortho-xylene.							
K095			1*	4	K095	B	100 (45.4)

Distillation bottoms from the production of 1,1,1-trichloroeth ane.							
K096			1*	4	K096	B	100 (45.4)
Heavy ends from the heavy ends column from the production of 1,1,1-trichloroeth ane.							
K097			1*	4	K097	X	1 (0.454)
Vacuum stripper discharge from the chlordane chlorinator in the production of chlordane.							
K098			1*	4	K098	X	1 (0.454)
Untreated process wastewater from the production of toxaphene.							
K099			1*	4	K099	A	10 (4.54)
Untreated wastewater from the production of 2,4-D.							
K100			1*	4	K100	A	10 (4.54)
Waste leaching solution from acid leaching of emission control dust/sludge from secondary lead smelting.							
K101			1*	4	K101	X	1 (0.454)

Distillation tar residues from the distillation of aniline-based compounds in the production of veterinary pharmaceuticals from arsenic or organo-arsenic compounds.							
K102			1*	4	K102	X	1 (0.454)
Residue from the use of activated carbon for decolorization in the production of veterinary pharmaceuticals from arsenic or organo-arsenic compounds.							
K103			1*	4	K103	B	100 (45.4)
Process residues from aniline extraction from the production of aniline.							
K104			1*	4	K104	A	10 (4.54)
Combined wastewater streams generated from nitrobenzene/aniline production.							
K105			1*	4	K105	A	10 (4.54)
Separated aqueous stream from the reactor product washing step in the production of chlorobenzenes.							
K106			1*	4	K106	X	1 (0.454)

Wastewater treatment sludge from the mercury cell process in chlorine production.							
K107			10	4	K107	X	10 (4.54)
Column bottoms from product separation from the production of 1,1-dimethylhydr azine (UDMH) from carboxylic acid hydrazines							
K108			10	4	K108	X	10 (4.54)
Condensed column overheads from product separation and condensed reactor vent gases from the production of 1,1-dimethylhydr azine (UDMH) from carboxylic acid hydrazides							
K109			10	4	K109	X	10 (4.54)
Spent filter cartridges from product purification from the production of 1,1-dimethylhydrazin e (UDMH) from carboxylic acid hydrazides							
K110			10	4	K110	X	10 (4.54)
Condensed column overheads from intermediate separation from the production of 1,1-dimethylhydr azine (UDMH) from carboxylic acid hydrazides							
K111			1*	4	K111	A	10 (4.54)

Product washwaters from the production of dinitrotoluene via nitration of toluene.							
K112			1*	4	K112	A	10 (4.54)
Reaction by-product water from the drying column in the production of toluenediamine via hydrogenation of dinitrotoluene.							
K113			1*	4	K113	A	10 (4.54)
Condensed liquid light ends from the purification of toluenediamine in the production of toluenediamine via hydrogenation of dinitrotoluene.							
K114			1*	4	K114	A	10 (4.54)
Vicinals from the purification of toluenediamine in the production of toluenediamine via hydrogenation of dinitrotoluene.							
K115			1*	4	K115	A	10 (4.54)
Heavy ends from the purification of toluenediamine in the production of toluenediamine via hydrogenation of dinitrotoluene.							
K116			1*	4	K116	A	10 (4.54)

Organic condensate from the solvent recovery column in the production of toluene diisocyanate via phosgenation of toluenediamine.							
K117			1*	4	K117	X	1 (0.454)
Wastewater from the reaction vent gas scrubber in the production of ethylene bromide via bromination of ethene.							
K118			1*	4	K118	X	1 (0.454)
Spent absorbent solids from purification of ethylene dibromide in the production of ethylene dibromide.							
K123			1*	4	K123	A	10 (4.54)
Process wastewater (including supernates, filtrates, and washwaters) from the production of ethylenebisdithiocarbamic acid and its salts.							
K124			1*	4	K124	A	10 (4.54)
Reactor vent scrubber water from the production of ethylenebisdithiocarbamic acid and its salts.							
K125			1*	4	K125	A	10 (4.54)

Filtration, evaporation, and centrifugation solids from the production of ethylenebisdithiocarbamic acid and its salts.							
K126			1*	4	K126	A	10 (4.54)
Baghouse dust and floor sweepings in milling and packaging operations from the production or formulation of ethylenebisdithiocarbamic acid and its salts.							
K131			100	4	K131	X	100 (45.4)
Wastewater from the reactor and spent sulfuric acid from the acid dryer in the production of methyl bromide.							
K132			1000	4	K132	X	1000 (454)
Spent absorbent and wastewater solids from the production of methyl bromide.							
K136			1*	4	K136	X	1 (0.454)
Still bottoms from the purification of ethylene dibromide in the production of ethylene dibromide via bromination of ethene.							
K141			1*	4	K141	X	1 (0.454)

Process related from the recovery of coal tar, including, but not limited to, tar collecting sump residues from the production of coke by-products produced from coal. This listing does not include K087 (decanter tank tar sludge from coking operations.)							
K142			1*	4	K142	X	1 (0.454)
Tar storage tank residues from the production of coke from coal or from the recovery of coke by-products produced from coal.							
K143			1*	4	K143	X	1 (0.454)
Process residues from the recovery of light oil, including, but not limited to, those generated in stills, decanters, and wash oil recovery units from the recovery of coke by-products produced from coal.							
K144			1*	4	K144	X	1 (0.454)
Wastewater sump residues from light oil refining, including, but not limited to, intercepting or contamination sump sludges from the recovery of coke by-products produced from coal.							

K145			1*	4	K145	X	1 (0.454)
Residues from naphthalene collection and recovery operations from the recovery of coke by-products produced from coal.							
K147			1*	4	K147	X	1 (0.454)
Tar storage tank residues from coal tar refining.							
K148			1*	4	K148	X	1 (0.454)
Residues from coal tar distillation, including, but not limited to, still bottoms.							
K149			1*	4	K149	A	10 (4.54)
Distillation bottoms from the production of alpha- (or methyl-) chlorinated toluenes, ring-chlorinated toluenes, benzoyl chlorides, and compounds with mixtures of these functional groups. [This waste does not include still bottoms from the distillation of benzyl chloride.							
K150			1*	4	K150	A	10 (4.54)

Organic residuals, excluding spent carbon adsorbent, from the spent chlorine gas and hydrochloric acid recovery processes associated with the production of alpha- (or methyl-) chlorinated toluenes, ring-chlorinated toluenes, benzoyl chlorides, and compounds with mixtures of these functional groups.							
K151			1*	4	K151	A	10 (4.54)
Wastewater treatment sludges, excluding neutralization and biological sludges, generated during the treatment of wastewaters from the production of alpha- (or methyl-) chlorinated toluenes, ring-chlorinated toluenes, benzoyl chlorides, and compounds with mixtures of these functional groups.							
Indicates the statutory source as defined by 1, 2, 3, and 4 below.							

No reporting of releases of this hazardous substance is required if the diameter of the pieces of the solid metal released is equal to or exceeds 100 micrometers (0.004 inches).							
The RQ for asbestos is limited to friable forms only.							
1-indicates that the statutory source for designation of this hazardous substance under CERCLA is CWA Section 311(b)(4).							
2-indicates that the statutory source for designation of this hazardous substance under CERCLA is CWA Section 307(a).							
3-indicates that the statutory source for designation of this hazardous substance under CERCLA is CAA Section 112.							
4-Indicates that the statutory source for designation of this hazardous substance under CERCLA is RCRA Section 3001.							
1*-indicates that the 1-pound RQ is a CERCLA statutory RQ.							

F1BBBB Indicates that the RQ is subject to change when the assessment of potential carcinogenicity is completed.							
F1BBBBF1 The Agency may adjust the statutory RQ for this hazardous substance in a future rulemaking; until then the statutory RQ applies.							

-The adjusted RQs for radionuclides may be found in appendix B to this table.

** -indicates that no RQ is being assigned to the generic or broad class.

Ci-Curie. The curie represents a rate of radioactive decay. One curie is the quantity of any radioactive nuclide which undergoes 3.7E 10 disintegrations per second.

Bq-Becquerel. The becquerel represents a rate of radioactive decay. One becquerel is the quantity of any radioactive nuclide which undergoes one disintegration per second. One curie is equal to 3.7E 10 becquerel.

@-Final RQs for all radionuclides apply to chemical compounds containing the radionuclides and elemental forms regardless of the diameter of pieces of solid material.

&-The adjusted RQ of one curie applies to all radionuclides not otherwise listed. Whenever the RQs in Table 302.4 and this appendix to the table are in conflict, the lowest RQ shall apply. For example, uranyl acetate and uranyl nitrate have adjusted RQs shown in Table 302.4 of 100 pounds, equivalent to about one-tenth the RQ level for uranium-238 listed in this appendix.

E-Exponent to the base 10. For example, 1.3E 2 is equal to 130 while 1.3E 3 is equal to 1300.

m-Signifies a nuclear isomer which is a radionuclide in a higher energy metastable state relative to the parent isotope.

f-Notification requirements for releases of mixtures or solutions of radionuclides can be found in Section 302.6(b) of this rule. Final RQs for the following four common radionuclide mixtures are provided: radium-226 in secular equilibrium with its daughters (0.053 curie); natural uranium (0.1 curie); natural uranium in secular equilibrium with its daughters (0.052 curie); and natural thorium in secular equilibrium with its daughters (0.011 curie).

[54 FR 33449, Aug. 14, 1989]

Editorial Note: For FEDERAL REGISTER citations affecting 302.4, see the List of CFR Sections Affected in the Finding Aids section of this volume.

V Limits for Air Contaminants

Substance	CAS No. (c)	ppm (a)[1]	mg/m^3 (b)[1]	Skin - Designation
Acetaldehyde	75-07-0	200	360	
Acetic acid	64-19-7	10	25	
Acetic anhydride	108-24-7	5	20	
Acetone	67-64-1	1000	2400	
Acetonitrile	75-05-8	40	70	
2-Acetylaminofluorine; see 1910.1014	53-96-3			
Acetylene dichloride; see 1,2-Dichloroethylene.				
Acetylene tetrabromide	79-27-6	1	14	
Acrolein	107-02-8	0.1	0.25	
Acrylamide	79-06-1		0.3	X
Acrylonitrile; see 1910.1045	107-13-1			
Aldrin	309-00-2		0.25	X
Allyl alcohol	107-18-6	2	5	X
Allyl chloride	107-05-1	1	3	
Allyl glycidyl ether (AGE)	106-92-3	(C)10	(C)45	
Allyl propyl disulfide	2179-59-1	2	12	
alpha-Alumina	1344-28-1			
Total dust			15	

Respirable fraction			5	
Aluminum, metal (as Al)	7429-90-5			
Total dust			15	
Respirable fraction			5	
4-Aminodiphenyl; see 1910.1011	92-67-1			
2-Aminoethanol; see Ethanolamine.				
2-Aminopyridine	504-29-0	0.5	2	
Ammonia	7664-41-7	50	35	
Ammonium sulfamate	7773-06-0			
Total dust			15	
Respirable fraction			5	
n-Amyl acetate	628-63-7	100	525	
sec-Amyl acetate	626-38-0	125	650	
Aniline and homologs	62-53-3	5	19	X
Anisidine (o-, p-isomers)	29191-52-4		0.5	X
Antimony and compounds (as Sb)	7440-36-0		0.5	
ANTU (alpha Naphthylthiourea)	86-88-4		0.3	
Arsenic, inorganic compounds (as As); see 1910.1018	7440-38-2			
Arsenic, organic compounds (as As)	7440-38-2		0.5	
Arsine	7784-42-1	0.05	0.2	
Asbestos; see 1910.1001	[4]			
Azinphos-methyl	86-50-0		0.2	X
Barium, soluble compounds (as Ba)	7440-39-3		0.5	
Barium sulfate	7727-43-7			
Total dust			15	
Respirable fraction			5	
Benomyl	17804-35-2			
Total dust			15	
Respirable fraction			5	

Benzene; see 1910.1028	71-43-2			
Benzidine; see 1910.1010	92-87-5			
p-Benzoquinone; see Quinone.				
Benzo(a)pyrene; see Coal tar pitch volatiles.				
Benzoyl peroxide	94-36-0		5	
Benzyl chloride	100-44-7	1	5	
Beryllium and beryllium compounds (as Be)	7440-41-7		[2]	
Biphenyl; see Diphenyl.				
Bismuth telluride, Undoped	1304-82-1			
Total dust			15	
Respirable fraction			5	
Boron oxide	1303-86-2			
Total dust			15	
Boron trifluoride	7637-07-2	(C)1	(C)3	
Bromine	7726-95-6	0.1	0.7	
Bromoform	75-25-2	0.5	5	X
Butadiene (1,3-Butadiene)	106-99-0	1000	2200	
Butanethiol; see Butyl mercaptan.				
2-Butanone (Methyl ethyl ketone)	78-93-3	200	590	
2-Butoxyethanol	111-76-2	50	240	X
n-Butyl-acetate	123-86-4	150	710	
sec-Butyl acetate	105-46-4	200	950	
tert-Butyl acetate	540-88-5	200	950	
n-Butyl alcohol	71-36-3	100	300	
sec-Butyl alcohol	78-92-2	150	450	
tert-Butyl alcohol	75-65-0	100	300	
Butylamine	109-73-9	(C)5	(C)15	X
tert-Butyl chromate (as CrO^3)	1189-85-1		(C)0.1	X
n-Butyl glycidyl ether (BGE)	2426-08-6	50	270	
Butyl mercaptan	109-79-5	10	35	

p-tert-Butyltoluene	98-51-1	10	60	
Cadmium (as Cd); see 1910.1027	7440-43-9			
Calcium carbonate	1317-65-3			
Total dust			15	
Respirable fraction			5	
Calcium hydroxide	1305-62-0			
Total dust			15	
Respirable fraction			5	
Calcium oxide	1305-78-8		5	
Calcium silicate	1344-95-2			
Total dust			15	
Respirable fraction			5	
Calcium sulfate	7778-18-9			
Total dust			15	
Respirable fraction			5	
Camphor, synthetic	76-22-2		2	
Carbaryl (Sevin)	63-25-2		5	
Carbon black	1333-86-4		3.5	
Carbon dioxide	124-38-9	5000	9000	
Carbon disulfide	75-15-0		[2]	
Carbon monoxide	630-08-0	50	55	
Carbon tetrachloride	56-23-5		[2]	
Cellulose	9004-34-6			
Total dust			15	
Respirable fraction			5	
Chlordane	57-74-9		0.5	X
Chlorinated camphene	8001-35-2		0.5	X
Chlorinated diphenyl oxide	55720-99-5		0.5	
Chlorine	7782-50-5	(C)1	(C)3	
Chlorine dioxide	10049-04-4	0.1	0.3	

Chlorine trifluoride	7790-91-2	(C)0.1	(C)0.4	
Chloroacetaldehyde	107-20-0	(C)1	(C)3	
a-Chloroacetophenone (Phenacyl chloride)	532-27-4	0.05	0.3	
Chlorobenzene	108-90-7	75	350	
o-Chlorobenzylidene malononitrile	2698-41-1	0.05	0.4	
Chlorobromomethane	74-97-5	200	1050	

2-Chloro-1,3-butadiene; see beta-Chloroprene.				
Chlorodiphenyl (42% Chlorine) (PCB)	53469-21-9		1	X
Chlorodiphenyl (54% Chlorine) (PCB)	11097-69-1		0.5	X

1-Chloro-2,3-epoxypropane; see Epichlorohydrin.				
2-Chloroethanol; see Ethylene chlorohydrin.				
Chloroethylene; see Vinyl chloride.				
Chloroform (Trichloromethane)	67-66-3	(C)50	(C)240	
bis(Chloromethyl) ether; see 1910.1008	542-88-1			
Chloromethyl methyl ether; see 1910.1006	107-30-2			
1-Chloro-1-nitropropane	600-25-9	20	100	
Chloropicrin	76-06-2	0.1	0.7	
beta-Chloroprene	126-99-8	25	90	X
2-Chloro-6-(trichloromethyl) pyridine	1929-82-4			
Total dust			15	
Respirable fraction			5	
Chromic acid and chromates (as CrO^3)	[4]		[2]	
Chromium (II) compounds.				
(as Cr)	7440-47-3		0.5	
Chromium (III) compounds.				
(as Cr)	7440-47-3		0.5	
Chromium metal and insol. salts (as Cr)	7440-47-3		1	
Chrysene; see Coal tar pitch volatiles.				

Clopidol	2971-90-6			
Total dust			15	
Respirable fraction			5	
Coal dust (less than 5% SiO^2), respirable fraction			[3]	
Coal dust (greater than or equal to 5% SiO^2), respirable fraction			[3]	
Coal tar pitch volatiles (benzene soluble fraction), anthracene, BaP, phenanthrene, acridine, chrysene pyrene	65966-93-2		0.2	
Cobalt metal, dust, and fume (as Co)	7440-48-4		0.1	
Coke oven emissions; see 1910.1029.				
Copper	7440-50-8			
Fume (as Cu)			0.1	
Dusts and mists (as Cu)			1	
Cotton dust[e]; see 1910.1043			1	
Crag herbicide (Sesone)	136-78-7			
Total dust			15	
Respirable fraction			5	
Cresol, all isomers	1319-77-3	5	22	X
Crotonaldehyde	123-73-9; <cr>4170-30-3	2	6	
Cumene	98-82-8	50	245	X
Cyanides (as CN)	[4]		5	
Cyclohexane	110-82-7	300	1050	
Cyclohexanol	108-93-0	50	200	
Cyclohexanone	108-94-1	50	200	
Cyclohexene	110-83-8	300	1015	
Cyclopentadiene	542-92-7	75	200	
2,4-D (Dichlorophenoxyacetic acid)	94-75-7		10	

Decaborane	17702-41-9	0.05	0.3	X
Demeton (Systox)	8065-48-3		0.1	X
Diacetone alcohol (4-Hydroxy-4-methyl-2-pentano ne)	123-42-2	50	240	
1,2-Diaminoethane; see Ethylenediamine.				
Diazomethane	334-88-3	0.2	0.4	
Diborane	19287-45-7	0.1	0.1	
1,2-Dibromo-3-chloropropane (CBCP); see1910.1044	96-12-8			
1,2-Dibromoethane; see Ethylene dibromide.				
Dibutyl phosphate	107-66-4	1	5	
Dibutyl phthalate	84-74-2		5	
o-Dichlorobenzene	95-50-1	(C)50	(C)300	
p-Dichlorobenzene	106-46-7	75	450	
3,3'-Dichlorobenzidine; see 1910.1007	91-94-1			
Dichlorodifluoromethane	75-71-8	1000	4950	
1,3-Dichloro-5,5-dimethyl hydantoin	118-52-5		0.2	
Dichlorodiphenyltrichloroetha ne (DDT)	50-29-3		1	X
1,1-Dichloroethane	75-34-3	100	400	
1,2-Dichloroethane; see Ethylene dichloride.				
1,2-Dichloroethylene	540-59-0	200	790	
Dichloroethyl ether	111-44-4	(C)15	(C)90	X
Dichloromethane; see Methylene chloride.				
Dichloromonofluoromethane	75-43-4	1000	4200	
1,1-Dichloro-1-nitroethane	594-72-9	(C)10	(C)60	
1,2-Dichloropropane; see Propylene dichloride.				
Dichlorotetrafluoroethane	76-14-2	1000	7000	
Dichlorvos (DDVP)	62-73-7		1	X
Dicyclopentadienyl iron	102-54-5			
Total dust			15	

Respirable fraction			5	
Dieldrin	60-57-1		0.25	X
Diethylamine	109-89-7	25	75	
2-Diethylaminoethanol	100-37-8	10	50	X
Diethyl ether; see Ethyl ether.				
Difluorodibromomethane	75-61-6	100	860	
Diglycidyl ether (DGE)	2238-07-5	(C)0.5	(C)2.8	
Dihydroxybenzene; see Hydroquinone.				
Diisobutyl ketone	108-83-8	50	290	
Diisopropylamine	108-18-9	5	20	X
4-Dimethylaminoazobenzene; see 1910.1015	60-11-7			
Dimethoxymethane; see Methylal.				
Dimethyl acetamide	127-19-5	10	35	X
Dimethylamine	124-40-3	10	18	
Dimethylaminobenzene; see Xylidine				
Dimethylaniline (N,N-Dimethylaniline)	121-69-7	5	25	X
Dimethylbenzene; see Xylene.				
Dimethyl-1,2-dibromo-2,2-dich loroethyl phosphate	300-76-5		3	
Dimethylformamide	68-12-2	10	30	X
2,6-Dimethyl-4-heptanone; see Diisobutyl ketone.				
1,1-Dimethylhydrazine	57-14-7	0.5	1	X
Dimethylphthalate	131-11-3		5	
Dimethyl sulfate	77-78-1	1	5	X
Dinitrobenzene (all isomers)			1	X
(ortho)	528-29-0			
(meta)	99-65-0			
(para)	100-25-4			
Dinitro-o-cresol	534-52-1		0.2	X
Dinitrotoluene	25321-14-6		1.5	X
Dioxane (Diethylene dioxide)	123-91-1	100	360	X

Diphenyl (Biphenyl)	92-52-4	0.2	1	
Diphenylmethane diisocyanate; see				
Methylene bisphenyl isocyanate.				
Dipropylene glycol methyl ether	34590-94-8	100	600	X
Di-sec octyl phthalate (Di-(2-ethylhexyl)				
phthalate)	117-81-7		5	
Emery	12415-34-8			
Total dust			15	
Respirable fraction			5	
Endosulfan	115-29-7		0.1	X
Endrin	72-20-8		0.1	X
Epichlorohydrin	106-89-8	5	19	X
EPN	2104-64-5		0.5	X
1,2-Epoxypropane; see Propylene oxide.				
2,3-Epoxy-1-propanol; see Glycidol.				
Ethanethiol; see Ethyl mercaptan.				
Ethanolamine	141-43-5	3	6	
2-Ethoxyethanol (Cellosolve)	110-80-5	200	740	X
2-Ethoxyethyl acetate (Cellosolve acetate)	111-15-9	100	540	X
Ethyl acetate	141-78-6	400	1400	
Ethyl acrylate	140-88-5	25	100	X
Ethyl alcohol (Ethanol)	64-17-5	1000	1900	
Ethylamine	75-04-7	10	18	
Ethyl amyl ketone (5-Methyl-3-heptanone)	541-85-5	25	130	
Ethyl benzene	100-41-4	100	435	
Ethyl bromide	74-96-4	200	890	
Ethyl butyl ketone (3-Heptanone)	106-35-4	50	230	
Ethyl chloride	75-00-3	1000	2600	
Ethyl ether	60-29-7	400	1200	
Ethyl formate	109-94-4	100	300	

Ethyl mercaptan	75-08-1	(C)10	(C)25	
Ethyl silicate	78-10-4	100	850	
Ethylene chlorohydrin	107-07-3	5	16	X
Ethylenediamine	107-15-3	10	25	
Ethylene dibromide	106-93-4		[2]	
Ethylene dichloride (1,2-Dichloroethane)	107-06-2		[2]	
Ethylene glycol dinitrate	628-96-6	(C)0.2	(C)1	X
Ethylene glycol methyl acetate; see Methyl cellosolve acetate.				
Ethyleneimine; see 1910.1012	151-56-4			
Ethylene oxide; see 1910.1047	75-21-8			
Ethylidene chloride; see 1,1-Dichloroethane.				
N-Ethylmorpholine	100-74-3	20	94	X
Ferbam	14484-64-1			
Total dust			15	
Ferrovanadium dust	12604-58-9		1	
Fluorides (as F)	[4]		2.5	
Fluorine	7782-41-4	0.1	0.2	
Fluorotrichloromethane (Trichlorofluoromethane)	75-69-4	1000	5600	
Formaldehyde; see 1910.1048	50-00-0			
Formic acid	64-18-6	5	9	
Furfural	98-01-1	5	20	X
Furfuryl alcohol	98-00-0	50	200	
Grain dust (oat, wheat, barley)			10	
Glycerin (mist)	56-81-5			
Total dust			15	
Respirable fraction			5	
Glycidol	556-52-5	50	150	
Glycol monoethyl ether; see 2-Ethoxyethanol.				

Graphite, natural, respirable dust	7782-42-5		[3]	
Graphite, synthetic				
Total dust			15	
Respirable fraction			5	
Guthion; see Azinphos methyl.				
Gypsum	13397-24-5			
Total dust			15	
Respirable fraction			5	
Hafnium	7440-58-6		0.5	
Heptachlor	76-44-8		0.5	X
Heptane (n-Heptane)	142-82-5	500	2000	
Hexachloroethane	67-72-1	1	10	X
Hexachloronaphthalene	1335-87-1		0.2	X
n-Hexane	110-54-3	500	1800	
2-Hexanone (Methyl n-butyl ketone)	591-78-6	100	410	
Hexone (Methyl isobutyl ketone)	108-10-1	100	410	
sec-Hexyl acetate	108-84-9	50	300	
Hydrazine	302-01-2	1	1.3	X
Hydrogen bromide	10035-10-6	3	10	
Hydrogen chloride	7647-01-0	(C)5	(C)7	
Hydrogen cyanide	74-90-8	10	11	X
Hydrogen fluoride (as F)	7664-39-3		[2]	
Hydrogen peroxide	7722-84-1	1	1.4	
Hydrogen selenide (as Se)	7783-07-5	0.05	0.2	
Hydrogen sulfide	7783-06-4		[2]	
Hydroquinone	123-31-9		2	
Iodine	7553-56-2	(C)0.1	(C)1	
Iron oxide fume	1309-37-1		10	
Isoamyl acetate	123-92-2	100	525	
Isoamyl alcohol (primary and secondary)	123-51-3	100	360	

Isobutyl acetate	110-19-0	150	700	
Isobutyl alcohol	78-83-1	100	300	
Isophorone	78-59-1	25	140	
Isopropyl acetate	108-21-4	250	950	
Isopropyl alcohol	67-63-0	400	980	
Isopropylamine	75-31-0	5	12	
Isopropyl ether	108-20-3	500	2100	
Isopropyl glycidyl ether (IGE)	4016-14-2	50	240	
Kaolin	1332-58-7			
Total dust			15	
Respirable fraction			5	
Ketene	463-51-4	0.5	0.9	
Lead, inorganic (as Pb); see 1910.1025	7439-92-1			
Limestone	1317-65-3			
Total dust			15	
Respirable fraction			5	
Lindane	58-89-9		0.5	X
Lithium hydride	7580-67-8		0.025	
L.P.G. (Liquefied petroleum gas)	68476-85-7	1000	1800	
Magnesite	546-93-0			
Total dust			15	
Respirable fraction			5	
Magnesium oxide fume	1309-48-4			
Total particulate			15	
Malathion	121-75-5			
Total dust			15	X
Maleic anhydride	108-31-6	0.25	1	
Manganese compounds (as Mn)	7439-96-5		(C)5	
Manganese fume (as Mn)	7439-96-5		(C)5	
Marble	1317-65-3			
Total dust			15	

Respirable fraction			5	
Mercury (aryl and inorganic) (as Hg)	7439-97-6		[2]	
Mercury (organo) alkyl compounds (as Hg)	7439-97-6		[2]	
Mercury (vapor) (as Hg)	7439-97-6		[2]	
Mesityl oxide	141-79-7	25	100	
Methanethiol; see Methyl mercaptan.				
Methoxychlor	72-43-5			
Total dust			15	
2-Methoxyethanol (Methyl cellosolve)	109-86-4	25	80	X
2-Methoxyethyl acetate (Methyl cellosolve acetate)	110-49-6	25	120	X
Methyl acetate	79-20-9	200	610	
Methyl acetylene (Propyne)	74-99-7	1000	1650	
Methyl acetylene-propadiene mixture (MAPP)		1000	1800	
Methyl acrylate	96-33-3	10	35	X
Methylal (Dimethoxy-methane)	109-87-5	1000	3100	
Methyl alcohol	67-56-1	200	260	
Methylamine	74-89-5	10	12	
Methyl amyl alcohol; see Methyl isobutyl carbinol.				
Methyl n-amyl ketone	110-43-0	100	465	
Methyl bromide	74-83-9	(C)20	(C)80	X
Methyl butyl ketone; see 2-Hexanone.				
Methyl cellosolve; see 2-Methoxyethanol.				
Methyl cellosolve acetate; see 2-Methoxyethyl acetate.				
Methyl chloride	74-87-3		[2]	
Methyl chloroform (1,1,1-Trichloroethane)	71-55-6	350	1900	
Methylcyclohexane	108-87-2	500	2000	
Methylcyclohexanol	25639-42-3	100	470	
o-Methylcyclohexanone	583-60-8	100	460	X
Methylene chloride	75-09-2		[2]	
Methyl ethyl ketone (MEK); see 2-Butanone.				
Methyl formate	107-31-3	100	250	

Methyl hydrazine (Monomethyl hydrazine)	60-34-4	(C)0.2	(C)0.35	X
Methyl iodide	74-88-4	5	28	X
Methyl isoamyl ketone	110-12-3	100	475	
Methyl isobutyl carbinol	108-11-2	25	100	X
Methyl isobutyl ketone; see Hexone.				
Methyl isocyanate	624-83-9	0.02	0.05	X
Methyl mercaptan	74-93-1	(C)10	(C)20	
Methyl methacrylate	80-62-6	100	410	
Methyl propyl ketone; see 2-Pentanone.				
alpha-Methyl styrene	98-83-9	(C)100	(C)480	
Methylene bisphenyl isocyanate (MDI)	101-68-8	(C)0.02	(C)0.2	
Mica; see Silicates.				
Molybdenum (as Mo)	7439-98-7			
Soluble compounds			5	
Insoluble compounds.				
Total dust			15	
Monomethyl aniline	100-61-8	2	9	X
Monomethyl hydrazine; see Methyl hydrazine.				
Morpholine	110-91-8	20	70	X
Naphtha (Coal tar)	8030-30-6	100	400	
Naphthalene	91-20-3	10	50	
alpha-Naphthylamine; see 1910.1004	134-32-7			
beta-Naphthylamine; see 1910.1009	91-59-8			
Nickel carbonyl (as Ni)	13463-39-3	0.001	0.007	
Nickel, metal and insoluble compounds (as Ni)	7440-02-0		1	
Nickel, soluble compounds (as Ni)	7440-02-0		1	
Nicotine	54-11-5		0.5	X
Nitric acid	7697-37-2	2	5	
Nitric oxide	10102-43-9	25	30	
p-Nitroaniline	100-01-6	1	6	X

Nitrobenzene	98-95-3	1	5	X
p-Nitrochlorobenzene	100-00-5		1	X
4-Nitrodiphenyl; see 1910.1003	92-93-3			
Nitroethane	79-24-3	100	310	
Nitrogen dioxide	10102-44-0	(C)5	(C)9	
Nitrogen trifluoride	7783-54-2	10	29	
Nitroglycerin	55-63-0	(C)0.2	(C)2	X
Nitromethane	75-52-5	100	250	
1-Nitropropane	108-03-2	25	90	
2-Nitropropane	79-46-9	25	90	
N-Nitrosodimethylamine; see 1910.1016.				
Nitrotoluene (all isomers)		5	30	X
o-isomer	88-72-2			
m-isomer	99-08-1			
p-isomer	99-99-0			
Nitrotrichloromethane; see Chloropicrin.				
Octachloronaphthalene	2234-13-1		0.1	X
Octane	111-65-9	500	2350	
Oil mist, mineral 8012-95-1	5			
Osmium tetroxide (as Os)	20816-12- 0		0.002	
Oxalic acid	144-62-7		1	
Oxygen difluoride	7783-41-7	0.05	0.1	
Ozone	10028-15-6	0.1	0.2	
Paraquat, respirable dust	4685-14-7 ; 1910-42-5; 2074-50-2		0.5	X
Parathion	56-38-2		0.1	X
Particulates not otherwise regulated (PNOR) [f].				
Total dust			15	
Respirable fraction			5	
PCB; see Chlorodiphenyl (42% and 54% chlorine).				
Pentaborane	19624-22-7	0.005	0.01	

Pentachloronaphthalene	1321-64-8		0.5	X
Pentachlorophenol	87-86-5		0.5	X
Pentaerythritol	115-77-5			
Total dust			15	
Respirable fraction			5	
Pentane	109-66-0	1000	2950	
2-Pentanone (Methyl propyl ketone)	107-87-9	200	700	
Perchloroethylene (Tetrachloroethylene)	127-18-4		[2]	
Perchloromethyl mercaptan	594-42-3	0.1	0.8	
Perchloryl fluoride	7616-94-6	3	13.5	
Perlite	93763-70-3			
Total dust			15	
Respirable fraction			5	
Petroleum distillates (Naphtha) (Rubber Solvent)		500	2000	
Phenol	108-95-2	5	19	X
p-Phenylene diamine	106-50-3		0.1	X
Phenyl ether, vapor	101-84-8	1	7	
Phenyl ether-biphenyl mixture, vapor		1	7	
Phenylethylene; see Styrene.				
Phenyl glycidyl ether (PGE)	122-60-1	10	60	
Phenylhydrazine	100-63-0	5	22	X
Phosdrin (Mevinphos)	7786-34-7		0.1	X
Phosgene (Carbonyl chloride)	75-44-5	0.1	0.4	
Phosphine	7803-51-2	0.3	0.4	
Phosphoric acid	7664-38-2		1	
Phosphorus (yellow)	7723-14-0		0.1	
Phosphorus pentachloride	10026-13-8		1	
Phosphorus pentasulfide	1314-80-3		1	
Phosphorus trichloride	7719-12-2	0.5	3	
Phthalic anhydride	85-44-9	2	12	

Picloram	1918-02-1			
Total dust			15	
Respirable fraction			5	
Picric acid	88-89-1		0.1	X
Pindone (2-Pivalyl-1,3-indandione)	83-26-1		0.1	
Plaster of Paris	26499-65-0			
Total dust			15	
Respirable fraction			5	
Platinum (as Pt)	7440-06-4			
Metal				
Soluble salts			0.002	
Portland cement	65997-15-1			
Total dust			15	
Respirable fraction			5	
Propane	74-98-6	1000	1800	
beta-Propriolactone; see 1910.1013	57-57-8			
n-Propyl acetate	109-60-4	200	840	
n-Propyl alcohol	71-23-8	200	500	
n-Propyl nitrate	627-13-4	25	110	
Propylene dichloride	78-87-5	75	350	
Propylene imine	75-55-8	2	5	X
Propylene oxide	75-56-9	100	240	
Propyne; see Methyl acetylene.				
Pyrethrum	8003-34-7		5	
Pyridine	110-86-1	5	15	
Quinone	106-51-4	0.1	0.4	
RDX; see Cyclonite.				
Rhodium (as Rh), metal fume and insoluble compounds	7440-16-6		0.1	
Rhodium (as Rh), soluble compounds	7440-16-6		0.001	
Ronnel	299-84-3		15	

Rotenone	83-79-4		5
Rouge			
Total dust			15
Respirable fraction			5
Selenium compounds (as Se)	7782-49-2		0.2
Selenium hexafluoride (as Se)	7783-79-1	0.05	0.4
Silica, amorphous, precipitated and gel	112926-00 -8		[3]
Silica, amorphous, diatomaceous earth, containing less than 1% crystalline silica	61790-53-2		[3]
Silica, crystalline cristobalite, respirable dust	14464-46-1		[3]
Silica, crystalline quartz, respirable dust	14808-60-7		[3]
Silica, crystalline tripoli (as quartz), respirable dust	1317-95-9		[3]
Silica, crystalline tridymite, respirable dust	15468-32-3		[3]
Silica, fused, respirable dust	60676-86-0		[3]
Silicates (less than 1% crystalline silica)			
Mica (respirable dust)	12001-26-2		[3]
Soapstone, total dust			[3]
Soapstone, respirable dust			[3]
Talc (containing asbestos); use asbestos limit; see 29 CFR 1910.1001			[3]
Talc (containing no asbestos), respirable dust	14807-96-6		[3]
Tremolite, asbestiform; see 1910.1001.			
Silicon	7440-21-3		
Total dust			15
Respirable fraction			5
Silicon carbide	409-21-2		
Total dust			15
Respirable fraction			5
Silver, metal and soluble compounds (as Ag)	7440-22-4		0.01

Soapstone; see Silicates.				
Sodium fluoroacetate	62-74-8		0.05	X
Sodium hydroxide	1310-73-2		2	
Starch	9005-25-8			
Total dust			15	
Respirable fraction			5	
Stibine	7803-52-3	0.1	0.5	
Stoddard solvent	8052-41-3	500	2900	
Strychnine	57-24-9		0.15	
Styrene	100-42-5		[2]	
Sucrose	57-50-1			
Total dust			15	
Respirable fraction			5	
Sulfur dioxide	7446-09-5	5	13	
Sulfur hexafluoride	2551-62-4	1000	6000	
Sulfuric acid	7664-93-9		1	
Sulfur monochloride	10025-67-9	1	6	
Sulfur pentafluoride	5714-22-7	0.025	0.25	
Sulfuryl fluoride	2699-79-8	5	20	
Systox; see Demeton.				
2,4,5-T (2,4,5-trichlorophenoxyacetic acid)	93-76-5		10	
Talc; see Silicates.				
Tantalum, metal and oxide dust	7440-25-7		5	
TEDP (Sulfotep)	3689-24-5		0.2	X
Tellurium and compounds (as Te)	13494-80-9		0.1	
Tellurium hexafluoride (as Te)	7783-80-4	0.02	0.2	
Temephos	3383-96-8			
Total dust			15	
Respirable fraction			5	
TEPP (Tetraethyl pyrophosphate)	107-49-3		0.05	X
Terphenyls	26140-60-3	(C)1	(C)9	

1,1,1,2-Tetrachloro-2,2-diflu oroethane	76-11-9	500	4170	
1,1,2,2-Tetrachloro-1,2-diflu oroethane	76-12-0	500	4170	
1,1,2,2-Tetrachloroethane	79-34-5	5	35	X
Tetrachloroethylene; see Perchloroethylene.				
Tetrachloromethane; see Carbon tetrachloride.				
Tetrachloronaphthalene	1335-88-2		2	X
Tetraethyl lead (as Pb)	78-00-2		0.075	X
Tetrahydrofuran	109-99-9	200	590	
Tetramethyl lead (as Pb)	75-74-1		0.075	X
Tetramethyl succinonitrile	3333-52-6	0.5	3	X
Tetranitromethane	509-14-8	1	8	
Tetryl (2,4,6-Trinitrophenylmethylni tramine)	479-45-8		1.5	X
Thallium, soluble compounds (as Tl)	7440-28-0		0.1	X
4,4'-Thiobis (6-tert, Butyl-m-cresol)	96-69-5			
Total dust			15	
Respirable fraction			5	
Thiram	137-26-8		5	
Tin, inorganic compounds (except oxides) (as Sn)	7440-31-5		2	
Tin, organic compounds (as Sn)	7440-31-5		0.1	
Titanium dioxide	13463-67-7			
Total dust			15	
Toluene	108-88-3		[2]	
Toluene-2,4-diisocyanate (TDI)	584-84-9	(C)0.02	(C)0.14	
o-Toluidine	95-53-4	5	22	X
Toxaphene; see Chlorinated camphene.				
Tremolite; see Silicates.				
Tributyl phosphate	126-73-8		5	
1,1,1-Trichloroethane; see Methyl chloroform.				
1,1,2-Trichloroethane	79-00-5	10	45	X
Trichloroethylene	79-01-6		[2]	

Trichloromethane; see Chloroform.				
Trichloronaphthalene	1321-65-9		5	X
1,2,3-Trichloropropane	96-18-4	50	300	
1,1,2-Trichloro-1,2,2-trifluo roethane	76-13-1	1000	7600	
Triethylamine	121-44-8	25	100	
Trifluorobromomethane	75-63-8	1000	6100	
2,4,6-Trinitrophenyl; see Picric acid.				
2,4,6-Trinitrophenylmethylnit ramine; see Tetryl.				
2,4,6-Trinitrotoluene (TNT)	118-96-7		1.5	X
Triorthocresyl phosphate	78-30-8		0.1	
Triphenyl phosphate	115-86-6		3	
Turpentine	8006-64-2	100	560	
Uranium (as U)	7440-61-1			
Soluble compounds			0.05	
Insoluble compounds			0.05	
Vanadium	1314-62-1			
Respirable dust (as V^2O^5)			(C)0.5	
Fume (as V^2O^5)			(C)0.1	
Vegetable oil mist				
Total dust			15	
Respirable fraction			5	
Vinyl benzene; see Styrene.				
Vinyl chloride; see 1910.1017	75-01-4			
Vinyl cyanide; see Acrylonitrile.				
Vinyl toluene	25013-15-4	100	480	
Warfarin	81-81-2		0.1	
Xylenes (o-, m-, p-isomers)	1330-20-7	100	435	
Xylidine	1300-73-8	5	25	X
Yttrium	7440-65-5		1	
Zinc chloride fume	7646-85-7		1	
Zinc oxide fume	1314-13-2		5	

Zinc oxide	1314-13-2	
Total dust		15
Respirable fraction		5
Zinc stearate	557-05-1	
Total dust		15
Respirable fraction		5
Zirconium compounds (as Zr)	7440-67-7	5

[1] The PELs are 8-hour TWAs unless otherwise noted; a (C) designation denotes a ceiling limit. They are to be determined from breathing-zone air samples.

(a) Parts of vapor or gas per million parts of contaminated air by volume at 25°C and 760 torr.

(b) Milligrams of substance per cubic meter of air. When entry is in this column only, the value is exact; when listed with a ppm entry, it is approximate.

(c) The CAS number is for information only. Enforcement is based on the substance name. For an entry covering more than one metal compound, measured as the metal, the CAS number for the metal is given, not CAS numbers for the individual compounds.

(d) The final benzene standard in 1910.1028 applies to all occupational exposures to benzene except in some circumstances the distribution and sale of fuels, sealed containers and pipelines, coke production, oil and gas drilling and production, natural gas processing, and the percentage exclusion for liquid mixtures; for the excepted subsegments, the benzene limits in Appndix VI apply. See 1910.1028 for specific circumstances.

(e) This 8-hour TWA applies to respirable dust as measured by a vertical elutriator cotton dust sampler or equivalent instrument. The time-weighted average applies to the cottom waste processing operations of waste recycling (sorting, blending, cleaning and willowing)and garnetting. See also 1910.1043 for cotton dust limits applicable to other sectors.

(f) All inert or nuisance dusts, whether mineral, inorganic, or organic, not listed specifically by substance name are covered by the Particulates Not Otherwise Regulated (PNOR) limit which is the same as the inert or nuisance dust limit of Appendix VII.

[2] See Appendix VI.

[3] See Appendix VII.

[4] Varies with compound.

VI Acceptable Ceiling Concentration

Substanc	8-hour time weighted average	Acceptable ceiling concentration	Acceptable maximum peak above above the acceptable ceiling concentration for an 8-hr shift	
			Concentrati on	Maximum duration
Benzene [a] (Z37.40-1969)	10 ppm	25 ppm	50 ppm	10 min
Beryllium and beryllium compounds (Z37.29-1970)	2 æg/m ^{3}	5 æg/m ^{3}	25 æg/m ^{3}	30 min
Cadmium fume [b] (Z37.5-1970)	0.1 mg/m ^{3}	0.3 mg/m ^{3}		
Cadmium dust [b] (Z37.5-1970)	0.2 mg/m ^{3}	0.6 mg/m ^{3}		
Carbon disulfide (Z37.3-1968)	20 ppm	30 ppm	100 ppm	30 min
Carbon tetrachloride (Z37.17-1967)	10 ppm	25 ppm	200 ppm	5 min in any 4 hr
Chromic acid and chromates (Z37.7-1971)		1 mg/10m ^{3}		
Ethylene dibromide (Z37.31-1970)	20 ppm	30 ppm	50 ppm	5 min

Ethylene dichloride (Z37.21-1969)	50 ppm	100 ppm	200 ppm	5 min in any 3 hr
Fluoride as dust (Z37.28-1969)	2.5 mg/m³			
Formaldehyde; see 1910.1048				
Hydrogen fluoride (Z37.28-1969)	3 ppm			
Hydrogen sulfide (Z37.2-1966)		20 ppm	50 ppm	10 min once, only if no other meas. exp. occurs.
Mercury (Z37.8-1971)		1 mg/10m³		
Methyl chloride (Z37.18-1969)	100 ppm	200 ppm	300 ppm	5 min in any 3 hr
Methylene chloride (Z37.23-1969)	500 ppm	1,000 ppm	2,000 ppm	5 min in any 2 hr
Organo (alkyl) mercury (Z37.30-1969)	0.01 mg/m³	0.04 mg/m³		
Styrene (Z37.15-1969)	100 ppm	200 ppm	600 ppm	5 min in any 3 hr
Tetrachloroethyle ne (Z37.22-1967)	100 ppm	200 ppm	300 ppm	5 min in any 3 hr
Toluene (Z37.12-1967)	200 ppm	300 ppm	500 ppm	10 min
Trichloroethylene (Z37.19-1967)	100 ppm	200 ppm	300 ppm	5 min in any 2 hr

[a] This standard applies to the industry segments exempt from the 1 ppm 8-hour TWA and 5 ppm STEL of the benzene standard at 1910.1028.

[b] This standard applies to any operations or sectors for which the Cadmium standard, 1910.1027, is stayed or otherwise not in effect.

VII

Mineral Dusts

Substance	mppcf [a]	mg/m^3
Silica:		
Crystalline	250 [b]	10 mg/m^3 [e]
Quartz (Respirable)	----------------	-------------
	$\%SiO^2+5$	$\% SiO^2 + 2$
		30 mg/m^3
Quartz (Total Dust)	....	-------------
		$\% SiO^2 + 2$
Cristobalite: Use 1/2 the value calculated from the count or mass formulae for quartz		
Tridymite: Use 1/2 the value calculated from the formulae for quartz		
Amorphous, including natural diatomaceous earth	20	80 mg/m^3 ------------- $\%SiO^2$
Silicates (less than 1% crystalline silica):		
Mica	20	

Soapstone	20	
Talc (not containing asbestos)	20 [c]	
Talc (containing asbestos) Use asbestos limit		
Tremolite, asbestiform (see 29 CFR 1910.1001)		
Portland cement	50	
Graphite (Natural)	15	
Coal Dust:		
Respirable fraction less than 5% SiO^2	....	2.4 mg/m³ ------------- %SiO²+2
Respirable fraction greater than 5% SiO^2	....	2.4 mg/m³ [e] ------------- %SiO² +2
Inert or Nuisance Dust: [d]		
Respirable fraction	15	5 mg/m³
Total dust	50	15 mg/m³

Note - Conversion factors - mppcf X 35.3 = million particles per cubic meter = particles per c.c.

[a] Millions of particles per cubic foot of air, based on impinger samples counted by light-field techniques.

[b] The percentage of crystalline silica in the formula is the amount determined from airborne samples, except in those instances in which other methods have been shown to be applicable.

[c] Containing >1% quartz; if 1% quartz or more, use quartz limit.

[e] All inert or nuisance dusts, whether mineral, inorganic, or organic, not listed specifically by substance name are covered by this limit which is the same as the Particulates Not Otherwise Regulated (PNOR) limit in Appendix V.

[e] Both concentration and percent quartz for the application of this limit are to be determined from the fraction passing a size-selector with the following characteristics:

Aerodynamic diameter (unit density sphere)	**Percent passing selector**
2	90
2.5	75
3.5	50
5.0	25
10	0

The measurements under this note refer to the use of an AEC (now NRC) instrument. The respirable fraction of coal dust is determined with an MRE; the figure corresponding to that of 2.4 mg/m^3 in the table for coal dust is 4.5 mg/m^3.

VIII List of Extremely Hazardous Substances

CAS No.	Chemical name	Notes	Reportable quantity (pounds)	Threshold planning quantity (pounds)
75-86-5	Acetone Cyanohydrin		10	1,000
1752-30-3	Acetone Thiosemicarbazide	e	1	1,000/10,000
107-02-8	Acrolein		1	500
79-06-1	Acrylamide	d, l	5,000	1,000/10,000
107-13-1	Acrylonitrile	d, l	100	10,000
814-68-6	Acrylyl Chloride	e, h	1	100
111-69-3	Adiponitrile	e, l	1	1,000
116-06-3	Aldicarb	c	1	100/10,000
309-00-2	Aldrin	d	1	500/10,000
107-18-6	Allyl Alcohol		100	1,000
107-11-9	Allylamine	e	1	500
20859-73-8	Aluminum Phosphide	b	100	500
54-62-6	Aminopterin	e	1	500/10,000

78-53-5	Amiton	e	1	500
3734-97-2	Amiton Oxalate	e	1	100/10,000
7664-41-7	Ammonia	l	100	500
300-62-9	Amphetamine	e	1	1,000
62-53-3	Aniline	d, l	5,000	1,000
88-05-1	Aniline, 2,4,6-Trimethyl-	e	1	500
7783-70-2	Antimony Pentafluoride	e	1	500
1397-94-0	Antimycin A	c, e	1	1,000/10,000
86-88-4	ANTU		100	500/10,000
1303-28-2	Arsenic pentoxide	d	1	100/10,000
1327-53-3	Arsenous oxide	d, h	1	100/10,000
7784-34-1	Arsenous trichloride	d	1	500
7784-42-1	Arsine	e	1	100
2642-71-9	Azinphos-Ethyl	e	1	100/10,000
86-50-0	Azinphos-Methyl		1	10/10,000
98-87-3	Benzal Chloride	d	5,000	500
98-16-8	Benzenamine, 3-(Trifluoromethyl)-	e	1	500
100-14-1	Benzene, 1-(Chloromethyl)-4-Nitro-	e	1	500/10,000
98-05-5	Benzenearsonic Acid	e	1	10/10,000
3615-21-2	Benzimidazole, 4,5-Dichloro-2-(Trifluoromethyl)-	e, g	1	500/10,000
98-07-7	Benzotrichloride	d	10	500
100-44-7	Benzyl Chloride	d	100	500
140-29-4	Benzyl Cyanide	e, h	1	500
15271-41-7	Bicyclo[2.2.1]Heptane-2-Carbonitrile, 5-Chloro-6-((((Methylamino)Carbonyl)Oxy)Imino)-, (1s-(1-alpha, 2-beta, 4-alpha, 5-alpha, 6E))-	e	1	500/10,000
534-07-6	Bis(Chloromethyl) Ketone	e	1	10/10,000
4044-65-9	Bitoscanate	e	1	500/10,000

10294-34-5	Boron Trichloride	e	1	500
7637-07-2	Boron Trifluoride	e	1	500
353-42-4	Boron Trifluoride Compound With Methyl Ether (1:1)	e	1	1,000
28772-56-7	Bromadiolone	e	1	100/10,000
7726-95-6	Bromine	e, l	1	500
1306-19-0	Cadmium Oxide	e	1	100/10,000
2223-93-0	Cadmium Stearate	c, e	1	1,000/10,0 00
7778-44-1	Calcium arsenate	d	1	500/10,000
8001-35-2	Camphechlor	d	1	500/10,000
56-25-7	Cantharidin	e	1	100/10,000
51-83-2	Carbachol Chloride	e	1	500/10,000
26419-73-8	Carbamic Acid, Methyl-, 0-(((2,4-Dimethyl-1, 3-Dithiolan-2-yl)Methylene)Amino)-	e	1	100/10,000
1563-66-2	Carbofuran		10	10/10,000
75-15-0	Carbon Disulfide	l	100	10,000
786-19-6	Carbophenothion	e	1	500
57-74-9	Chlordane	d	1	1,000
470-90-6	Chlorfenvinfos	e	1	500
7782-50-5	Chlorine	10	100	
24934-91-6	Chlormephos	e	1	500
999-81-5	Chlormequat Chloride	e, h	1	100/10,000
79-11-8	Chloroacetic Acid	e	1	100/10,000
107-07-3	Chloroethanol	e	1	500
627-11-2	Chloroethyl Chloroformate	e	1	1,000
67-66-3	Chloroform	d, 1	10	10,000
542-88-1	Chloromethyl ether	d, h	10	100
107-30-2	Chloromethyl methyl ether	e, d	10	100
3691-35-8	Chlorophacinone	e	1	100/10,000
1982-47-4	Chloroxuron	e	1	500/10,000
21923-23-9	Chlorthiophos	e, h	1	500
10025-73-7	Chromic Chloride	e	1	1/10,000

62207-76-5	Cobalt, ((2,2'-(1,2-Ethanediylbis (Nitrilomethylidyne))Bis(6-Fluorophenolato))(2-)-N,N',O,O')-,	e	1	100/10,000
10210-68-1	Cobalt Carbonyl	e, h	1	10/10,000
64-86-8	Colchicine	e, h	1	10/10,000
56-72-4	Coumaphos		10	100/10,000
5836-29-3	Coumatetralyl	e	1	500/10,000
95-48-7	Cresol, o-	d	1,000	1,000/10,000
535-89-7	Crimidine	e	1	100/10,000
4170-30-3	Crotonaldehyde		100	1,000
123-73-9	Crotonaldehyde, (E)-		100	1,000
506-68-3	Cyanogen Bromide		1,000	500/10,000
506-78-5	Cyanogen Iodide	e	1	1,000/10,000
2636-26-2	Cyanophos	e	1	1,000
675-14-9	Cyanuric Fluoride	e	1	100
66-81-9	Cycloheximide	e	1	100/10,000
108-91-8	Cyclohexylamine	e, l	1	10,000
17702-41-9	Decaborane(14)	e	1	500/10,000
8065-48-3	Demeton	e	1	500
919-86-8	Demeton-S-Methyl	e	1	500
10311-84-9	Dialifor	e	1	100/10,000
19287-45-7	Diborane	e	1	100
111-44-4	Dichloroethyl ether	d	10	10,000
149-74-6	Dichloromethylphenylsilane	e	1	1,000
62-73-7	Dichlorvos		10	1,000
141-66-2	Dicrotophos	e	1	100
1464-53-5	Diepoxybutane	d	10	500
814-49-3	Diethyl Chlorophospate	e, h	1	500
1642-54-2	Diethylcarbamazine Citrate	e	1	100/10,000

71-63-6	Digitoxin	c, e	1	100/10,000
2238-07-5	Diglycidyl Ether	e	1	1,000
20830-75-5	Digoxin	e, h	1	10/10,000
115-26-4	Dimefox	e	1	500
60-51-5	Dimethoate	10		500/10,000
2524-03-0	Dimethyl Phosphorochloridothioate	e	1	500
77-78-1	Dimethyl sulfate	d	100	500
75-78-5	Dimethyldichlorosilane	e, h	1	500
57-14-7	Dimethylhydrazine	d	10	1,000
99-98-9	Dimethyl-p-Phenylenediamine	e	1	10/10,000
644-64-4	Dimetilan	e	1	500/10,000
534-52-1	Dinitrocresol		10	10/10,000
88-85-7	Dinoseb		1,000	100/10,000
1420-07-1	Dinoterb	e	1	500/10,000
78-34-2	Dioxathion	e	1	500
82-66-6	Diphacinone	e	1	10/10,000
152-16-9	Diphosphoramide, Octamethyl-		100	100
298-04-4	Disulfoton		1	500
514-73-8	Dithiazanine Iodide	e	1	500/10,000
541-53-7	Dithiobiuret		100	100/10,000
316-42-7	Emetine, Dihydrochloride	e, h	1	1/10,000
115-29-7	Endosulfan		1	10/10,000
2778-04-3	Endothion	e	1	500/10,000
72-20-8	Endrin		1	500/10,000
106-89-8	Epichlorohydrin	d,1	100	1,000
2104-64-5	EPN	e	1	100/10,000
50-14-6	Ergocalciferol	c, e	1	1,000/10,000

379-79-3	Ergotamine Tartrate	e	1	500/10,000
1622-32-8	Ethanesulfonyl Chloride, 2-Chloro-	e	1	500
10140-87-1	Ethanol, 1.2-Dichloro-, Acetate	e	1	1,000
563-12-2	Ethion		10	1,000
13194-48-4	Ethoprophos	e	1	1,000
538-07-8	Ethylbis(2-Chloroethyl)Amine	e, h	1	500
371-62-0	Ethylene Fluorohydrin	c, e, h	1	10
75-21-8	Ethylene oxide	d,l	10	1,000
107-15-3	Ethylenediamine		5,000	10,000
151-56-4	Ethyleneimine	d	1	500
542-90-5	Ethylthiocyanate	e	1	10,000
22224-92-6	Fenamiphos	e	1	10/10,000
122-14-5	Fenitrothion	e	1	500
115-90-2	Fensulfothion	e, h	1	500
4301-50-2	Fluenetil	e	1	100/10,000
7782-41-4	Fluorine	k	10	500
640-19-7	Fluoroacetamide	j	100	100/10,000
144-49-0	Fluoroacetic Acid	e	1	10/10,000
359-06-8	Fluoroacetyl Chloride	c, e	1	10
51-21-8	Fluorouracil	e	1	500/10,000
944-22-9	Fonofos	e	1	500
50-00-0	Formaldehyde	d,l	100	500
107-16-4	Formaldehyde Cyanohydrin	e, h	1	1,000
23422-53-9	Formetanate Hydrochloride	e, h	1	500/10,000
2540-82-1	Formothion	e	1	100
17702-57-7	Formparanate	e	1	100/10,000
21548-32-3	Fosthietan	e	1	500
3878-19-1	Fuberidazole	e	1	100/10,000
110-00-9	Furan		100	500
13450-90-3	Gallium Trichloride	e	1	500/10,000
77-47-4	Hexachlorocyclopentadiene	d, h	10	100

4835-11-4	Hexamethylenediamine, N,N'-Dibutyl-	e	1	500
302-01-2	Hydrazine	d	1	1,000
74-90-8	Hydrocyanic Acid		10	100
7647-01-0	Hydrogen chloride (gas only)	e,1	5,000	500
7664-39-3	Hydrogen Fluoride		100	100
7722-84-1	Hydrogen Peroxide (Conc >52%)	e, l	1	1,000
7783-07-5	Hydrogen Selenide	e	1	10
7783-06-4	Hydrogen Sulfide	l	100	500
123-31-9	Hydroquinone	l	1	500/10,000
13463-40-6	Iron, Pentacarbonyl-	e	1	100
297-78-9	Isobenzan	e	1	100/10,000
78-82-0	Isobutyronitrile	e, h	1	1,000
102-36-3	Isocyanic Acid, 3,4-Dichlorophenyl Ester	e	1	500/10,000
465-73-6	Isodrin	1		100/10,000
55-91-4	Isofluorphate	c	100	100
4098-71-9	Isophorone Diisocyanate	b, e	1	100
108-23-6	Isopropyl Chloroformate	e	1	1,000
119-38-0	Isoproplymethylpyrazolyl Dimethylcarbamate	e	1	500
78-97-7	Lactonitrile	e	1	1,000
21609-90-5	Leptophos	e	1	500/10,000
541-25-3	Lewisite	c, e, h	1	10
58-89-9	Lindane	d	1	1,000/10,000
7580-67-8	Lithium Hydride	b, e	1	100
109-77-3	Malononitrile		1,000	500/10,000
12108-13-3	Manganese, Tricarbonyl Methylcyclopentadienyl	e, h	1	100
51-75-2	Mechlorethamine	c, e	1	10
950-10-7	Mephosfolan	e	1	500
1600-27-7	Mercuric Acetate	e	1	500/10,000
7487-94-7	Mercuric Chloride	e	1	500/10,000
21908-53-2	Mercuric Oxide	e	1	500/10,000

10476-95-6	Methacrolein Diacetate	e	1	1,000
760-93-0	Methacrylic Anhydride	e	1	500
126-98-7	Methacrylonitrile	h	1,000	500
920-46-7	Methacryloyl Chloride	e	1	100
30674-80-7	Methacryloyloxyethyl Isocyanate	e, h	1	100
10265-92-6	Methamidophos	e	1	100/10,000
558-25-8	Methanesulfonyl Fluoride	e	1	1,000
950-37-8	Methidathion	e	1	500/10,000
2032-65-7	Methiocarb		10	500/10,000
16752-77-5	Methomyl	h	100	500/10,000
151-38-2	Methoxyethylmercuric Acetate	e	1	500/10,000
80-63-7	Methyl 2-Chloroacrylate	e	1	500
74-83-9	Methyl Bromide	l	1,000	1,000
79-22-1	Methyl Chloroformate	d, h	1,000	500
60-34-4	Methyl Hydrazine		10	500
624-83-9	Methyl Isocyanate		10	500
556-61-6	Methyl Isothiocyanate	b, e	1	500
74-93-1	Methyl Mercaptan	l	100	500
3735-23-7	Methyl Phenkapton	e	1	500
676-97-1	Methyl Phosphonic Dichloride	b, e	1	100
556-64-9	Methyl Thiocyanate	e	1	10,000
78-94-4	Methyl Vinyl Ketone	e	1	10
502-39-6	Methylmercuric Dicyanamide	e	1	500/10,000
75-79-6	Methyltrichlorosilane	e, h	1	500
1129-41-5	Metolcarb	e	1	100/10,000
7786-34-7	Mevinphos		10	500
315-18-4	Mexacarbate		1,000	500/10,000
50-07-7	Mitomycin C	d	10	500/10,000
6923-22-4	Monocrotophos	e	1	10/10,000
2763-94-4	Muscimol		1,000	500/10,000
505-60-2	Mustard Gas	e, h	1	500
13463-39-3	Nickel carbonyl	d	10	1
54-11-5	Nicotine	c	100	100

65-30-5	Nicotine sulfate	e	100	100/10,000
7697-37-2	Nitric Acid		1,000	1,000
10102-43-9	Nitric Oxide	c	10	100
98-95-3	Nitrobenzene	l	1,000	10,000
1122-60-7	Nitrocyclohexane	e	1	500
10102-44-0	Nitrogen Dioxide		10	100
62-75-9	Nitrosodimethylamine	d ,h	10	1,000
991-42-4	Norbormide	e	1	100/10,000
0	Organorhodium Complex (PMN-82-147)	e	1	10/10,000
630-60-4	Ouabain	c, e	1	100/10,000
23135-22-0	Oxamyl	e	1	100/10,000
78-71-7	Oxetane, 3,3-Bis(Chloromethyl)-	1	e	500
2497-07-6	Oxydisulfoton	e, h	1	500
10028-15-6	Ozone	e	1	100
1910-42-5	Paraquat	e	1	10/10,000
2074-50-2	Paraquat Methosulfate	e	1	10/10,000
56-38-2	Parathion	c, d	10	100
298-00-0	Parathion-Methyl	c	100	100/10,000
12002-03-8	Paris Green	d	1	500/10,000
19624-22-7	Pentaborane	e	1	500
2570-26-5	Pentadecylamine	e	1	100/10,000
79-21-0	Peracetic Acid	e	1	500
594-42-3	Perchloromethylmercaptan		100	500
108-95-2	Phenol		1,000	500/10,000
4418-66-0	Phenol, 2,2'-Thiobis(4-Chloro-6-Methyl)-	e	1	100/10,000
64-00-6	Phenol, 3-(1-Methylethyl)-, Methylcarbamate	e	1	500/10,000
58-36-6	Phenoxarsine, 10,10'-Oxydi-	e	1	500/10,000
696-28-6	Phenyl Dichloroarsine	d, h	1	500
59-88-1	Phenylhydrazine Hydrochloride	e	1	1,000/10,000

62-38-4	Phenylmercury Acetate		100	500/10,000
2097-19-0	Phenylsilatrane	e, h	1	100/10,000
103-85-5	Phenylthiourma		108	100/10,000
298-02-2	Phorate		10	10
4104-14-7	Phosacetim	e	1	100/10,000
947-02-4	Phosfolan	e	1	100/10,000
75-44-5	Phosgene	l	10	10
732-11-6	Phosmet	e	1	10/10,000
13171-21-6	Phosphamidon	e	1	100
7803-51-2	Phosphine		100	500
2703-13-1	Phosphonothioic Acid, Methyl-, O-Ethyl O-(4-(Methylthio)Phenyl) Ester	e	1	500
50782-69-9	Phosphonothioic Acid, Methyl-, S-(2-(Bis(1-Methylethyl)Amino)Ethyl O-Ethyl Ester	e	1	100
2665-30-7	Phosphonothioic Acid, Methyl-, O-(4-Nitrophenyl) O-Phenyl Ester	e	1	500
3254-63-5	Phosphoric Acid, Dimethyl 4-(Methylthio) Phenyl Ester	e	1	500
2587-90-8	Phosphorothioic Acid, O,O-Dimethyl-S-(2-Methylthio) Ethyl Ester	c, e, g	1	500
7723-14-0	Phosphorus	b, h	1	100
10025-87-3	Phosphorus Oxychloride	d	1,000	500
10026-13-8	Phosphorus Pentachloride	b, e	1	500
1314-56-3	Phosphorus Pentoxide	b, e	1	10
7719-12-2	Phosphorus Trichloride		1,000	1,000
57-47-6	Physostigmine	e	1	100/10,000
57-64-7	Physostigmine, Salicylate (1:1)	e	1	100/10,000
124-87-8	Picrotoxin	e	1	500/10,000
110-89-4	Piperidine	e	1	1,000
23505-41-1	Pirimifos-Ethyl	e	1	1,000
10124-50-2	Potassium arsenite	d	1	500/10,000

151-50-8	Potassium Cyanide	b	10	100
506-61-6	Potassium Silver Cyanide	b	1	500
2631-37-0	Promecarb	e, h	1	500/10,000
106-96-7	Propargyl Bromide	e	1	10
57-57-8	Propiolactone, Beta-	e	1	500
107-12-0	Propionitrile		10	500
542-76-7	Propionitrile, 3-Chloro-		1,000	1,000
70-69-9	Propiophenone, 4-Amino-	e, g	1	100/10,000
109-61-5	Propyl Chloroformate	e	1	500
75-56-9	Propylene Oxide	l	100	10,000
75-55-8	Propyleneimine	d	1	10,000
2275-18-5	Prothoate	e	1	100/10,000
129-00-0	Pyrene	c	5,000	1,000/10,000
140-76-1	Pyridine, 2-Methyl-5-Vinyl-	e	1	500
504-24-5	Pyridine, 4-Amino-	h	1,000	500/10,000
1124-33-0	Pyridine, 4-Nitro-, 1-Oxide	e	1	500/10,000
53558-25-1	Pyriminil	e, h	1	100/10,000
14167-18-1	Salcomine	e	1	500/10,000
107-44-8	Sarin	e, h	1	10
7783-00-8	Selenious Acid		10	1,000/10,000
7791-23-3	Selenium Oxychloride	e	1	500
563-41-7	Semicarbazide Hydrochloride	e	1	1,000/10,000
3037-72-7	Silane, (4-Aminobutyl)Diethoxymethyl-	e	1	1,000
7631-89-2	Sodium Arsenate	d	1,000	1,000/10,000
7784-46-5	Sodium arsenite	d	1	500/10,000
26628-22-8	Sodium Azide (Na(N3))	b	1,000	500
124-65-2	Sodium Cacodylate	e	1	100/10,000

143-33-9	Sodium Cyanide (Na(CN))	b	10	100
62-74-8	Sodium Fluoroacetate		10	10/10,000
13410-01-0	Sodium Selenate	e	1	100/10,000
10102-18-8	Sodium Selenite	h	100	100/10,000
10102-20-2	Sodium Tellurite	e	1	500/10,000
900-95-8	Stannane, Acetoxytriphenyl-	e, g	1	500/10,000
57-24-9	Strychnine	c	10	100/10,000
60-41-3	Strychnine sulfate	e	10	100/10,000
3689-24-5	Sulfotep		100	500
3569-57-1	Sulfoxide, 3-Chloropropyl Octyl	e	1	500
7446-09-5	Sulfur Dioxide	e, l	1	500
7783-60-0	Sulfur Tetrafluoride	e	1	100
7446-11-9	Sulfur Trioxide	b, e	1	100
7664-93-9	Sulfuric Acid		1,000	1,000
77-81-6	Tabun	c, e, h	1	10
13494-80-9	Tellurium	e	1	500/10,000
7783-80-4	Tellurium Hexafluoride	e, k	1	100
107-49-3	TEPP		10	100
13071-79-9	Terbufos	e, h	1	100
78-00-2	Tetraethyllead	c, d	10	100
597-64-8	Tetraethyltin	c, e	1	100
75-74-1	Tetramethyllead	c, e, l	1	100
509-14-8	Tetranitromethane		10	500
10031-59-1	Thallium Sulfate	h	100	100/10,000
6533-73-9	Thallous Carbonate	c, h	100	100/10,000
7791-12-0	Thallous Chloride	c, h	100	100/10,000

2757-18-8	Thallous Malonate	c, e, h	1	100/10,000
7446-18-6	Thallous Sulfate		100	100/10,000
2231-57-4	Thiocarbazide	e	1	1,000/10,000
39196-18-4	Thiofanox		100	100/10,000
297-97-2	Thionazin		100	500
108-98-5	Thiophenol		100	500
79-19-6	Thiosemicarbazide		100	100/10,000
5344-82-1	Thiourea, (2-Chlorophenyl)-		100	100/10,000
614-78-8	Thiourea, (2-Methylphenyl)-	e	1	500/10,000
7550-45-0	Titanium Tetrachloride	e	1	100
584-84-9	Toluene 2,4-Diisocyanate		100	500
91-08-7	Toluene 2,6-Diisocyanate		100	100
110-57-6	Trans-1,4-Dichlorobutene	e	1	500
1031-47-6	Triamiphos	e	1	500/10,000
24017-47-8	Triazofos	e	1	500
76-02-8	Trichloroacety Chloride	e	1	500
115-21-9	Trichloroethylsilane	e, h	1	500
327-98-0	Trichloronate	e, k	1	500
98-13-5	Trichlorophenylsilane	e, h	1	500
1558-25-4	Trichloro(Chloromethyl)Silane	e	1	100
27137-85-5	Trichloro(Dichlorophenyl)Silane	e	1	500
998-30-1	Triethoxysilane	e	1	500
75-77-4	Trimethylchlorosilane	e	1	1,000
824-11-3	Trimethylolpropane Phosphite	e, h	1	100/10,000

1066-45-1	Trimethyltin Chloride	e	1	500/10,000
639-58-7	Triphenyltin Chloride	e	1	500/10,000
555-77-1	Tris(2-Chloroethyl)Amine	e, h	1	100
2001-95-8	Valinomycin	c, e	1	1,000/10,000
1314-62-1	Vanadium Pentoxide		1,000	100/10,000
108-05-4	Vinyl Acetate Monomer	d, l	5,000	1,000
81-81-2	Warfarin		100	500/10,000
129-06-6	Warfarin sodium	e, h	100	100/10,000
28347-13-9	Xylylene Dichloride	e	1	100/10,000
58270-08-9	Zinc, Dichloro(4,4-Dimethyl-5((((Methylamino) Carbonyl)Oxy)Imino)Pentanenitrile)-,(T-4)-	e	1	100/10,000
1314-84-7	Zinc Phosphide	b	100	500

*Only the statutory or final RQ is shown. For more information, see 40 CFR Table 302.4

Notes:

a This chemical does not meet acute toxicity criteria. Its TPQ is set at 10,000 pounds.

b This material is a reactive solid. The TPQ does not default to 10,000 pounds for non-powder, non-molten, non-solution form.

c The calculated TPQ changed after technical review as described in the technical support document.

d Indicates that the RQ is subject to change when the assessment of potential carcinogenicity and/or other toxicity is completed.

e Statutory reportable quantity for purposes of notification under SARA sect 304(a)(2).
f [Removed]
g New chemicals added that were not part of the original list of 402 substances.
h Revised TPQ based on new or re-evaluated toxicity data.
j TPQ is revised to its calculated value and does not change due to technical review as in proposed rule.
k The TPQ was revised after proposal due to calculation error.
l Chemicals on the original list that do not meet toxicity criteria but because of their high production volume and recognized toxicity are considered chemicals of concern (Other chemicals). [52 FR 13395, Apr. 22, 1987; 52 FR 15321, 15412, Apr. 28, 1987; 52 FR 48073-48074, Dec. 17, 1987; 53 FR 5575, Feb. 25, 1988; 54 FR 43165, Oct. 20, 1989; 54 FR 53063, Dec. 27, 1989; 55 FR 5546, Feb. 15, 1990]

INDEX

C

D

I

J

N

O

P

R

S

T